CLEP College Mathematics Workbook

Essential Learning Math Skills Plus Two College Math Practice Tests

By

Michael Smith & Reza Nazari

CLEP College Mathematics Workbook

Published in the United State of America By

The Math Notion

Web: WWW.MathNotion.Com

Email: info@Mathnotion.com

Copyright © 2020 by the Math Notion. All rights reserved. No part of this publication may be reproduced, stored in a retrieval system, or transmitted in any form or by any means, electronic, mechanical, photocopying, recording, scanning, or otherwise, except as permitted under Section 107 or 108 of the 1976 United States Copyright Ac, without permission of the author.

All inquiries should be addressed to the Math Notion.

ISBN: 978-1-63620-003-3

About the Author

Michael Smith has been a math instructor for over a decade now. He holds a master's degree in Management. Since 2006, Michael has devoted his time to both teaching and developing exceptional math learning materials. As a Math instructor and test prep expert, Michael has worked with thousands of students. He has used the feedback of his students to develop a unique study program that can be used by students to drastically improve their math score fast and effectively.

- **– SAT Math Practice Book**
- **– ACT Math Practice Book**
- **– PSAT Math Practice Book**
- **– Algebra Math Practice Books**
- **– Common Core Math Practice Books**
- **–many Math Education Workbooks, Exercise Books and Study Guides**

As an experienced Math teacher, Mr. Smith employs a variety of formats to help students achieve their goals: He tutors online and in person, he teaches students in large groups, and he provides training materials and textbooks through his website and through Amazon.

You can contact Michael via email at:

info@Mathnotion.com

Prepare for the CLEP College Mathematics with a Perfect Workbook!

CLEP College Mathematics Workbook is a learning workbook to prevent learning loss. It helps you retain and strengthen your Math skills and provides a strong foundation for success. This CLEP College Mathematics book provides you with a solid foundation to get ahead starts on your upcoming CLEP College Mathematics Test.

CLEP College Mathematics Workbook is designed by top math instructors to help students prepare for the CLEP College Mathematics course. It provides students with an in-depth focus on the CLEP College Mathematics concepts. This is a prestigious resource for those who need extra practice to succeed on the CLEP College Mathematics test.

CLEP College Mathematics Workbook contains many exciting and unique features to help you score higher on the CLEP College Mathematics test, including:

- Over 2,500 CLEP College Mathematics Practice questions with answers
- Complete coverage of all Math concepts which students will need to ace the CLEP College Mathematics test
- Two CLEP College Mathematics practice tests with detailed answers
- Content 100% aligned with the latest CLEP College Mathematics courses

This Comprehensive Workbook for the CLEP College Mathematics is a perfect resource for those CLEP College Math takers who want to review core content areas, brush-up in math, discover their strengths and weaknesses, and achieve their best scores on the CLEP College Mathematics test.

WWW.MathNotion.COM

... So Much More Online!

- ✓ FREE Math Lessons

- ✓ More Math Learning Books!

- ✓ Mathematics Worksheets

- ✓ Online Math Tutors

For a PDF Version of This Book

Please Visit WWW.MathNotion.com

contents

Chapter 1: Integers and Number Theory 13
   Adding and Subtracting Integers 14
   Multiplying and Dividing Integers 15
   Order of Operations 16
   Ordering Integers and Numbers 17
   Integers and Absolute Value 18
   Factoring Numbers 19
   Greatest Common Factor 20
   Least Common Multiple 21
   Sets 22
   Answers of Worksheets – Chapter 1 23

Chapter 2: Fractions and Decimals 27
   Simplifying Fractions 28
   Adding and Subtracting Fractions 29
   Multiplying and Dividing Fractions 30
   Adding and Subtracting Mixed Numbers 31
   Multiplying and Dividing Mixed Numbers 32
   Adding and Subtracting Decimals 33
   Multiplying and Dividing Decimals 34
   Comparing Decimals 35
   Rounding Decimals 36
   Answers of Worksheets – Chapter 2 37

Chapter 3: Proportions, Ratios, and Percent 41
   Simplifying Ratios 42
   Proportional Ratios 43
   Similarity and Ratios 44
   Ratio and Rates Word Problems 45
   Percentage Calculations 46
   Percent Problems 47

Discount, Tax and Tip 48
Percent of Change... 49
Simple Interest 50
Answers of Worksheets – Chapter 3 51

Chapter 4: Exponents... 55
Multiplication Property of Exponents 56
Zero and Negative Exponents.. 57
Division Property of Exponents 58
Powers of Products and Quotients 59
Negative Exponents and Negative Bases 60
Scientific Notation 61
Answers of Worksheets – Chapter 4 62

Chapter 5: Radicals Expressions... 65
Square Roots 66
Simplifying Radical Expressions 67
Multiplying Radical Expressions 68
Simplifying Radical Expressions Involving Fractions.............................. 69
Adding and Subtracting Radical Expressions... 70
Answers of Worksheets – Chapter 5 71

Chapter 6: Algebraic Expressions 73
Simplifying Variable Expressions... 74
Simplifying Polynomial Expressions.. 75
Translate Phrases into an Algebraic Statement 76
The Distributive Property.. 77
Evaluating One Variable Expressions 78
Evaluating Two Variables Expressions.. 79
Combining like Terms... 80
Answers of Worksheets – Chapter 6 81

Chapter 7: Equations and Inequalities.. 83
One–Step Equations 84
Multi–Step Equations 85
Graphing Single–Variable Inequalities 86
One–Step Inequalities 87

Multi-Step Inequalities 88
Systems of Equations 89
Systems of Equations Word Problems 90
Finding Midpoint 91
Finding Distance of Two Points 92
Answers of Worksheets – Chapter 7 93

Chapter 8: Linear Functions 97
Relation and Functions 98
Finding Slope 99
Graphing Lines Using Line Equation 100
Writing Linear Equations 101
Graphing Linear Inequalities 102
Write an Equation from a Graph 103
Rate of change 104
x and y intercepts 104
Slope–intercept Form 105
Point–slope Form 106
Graphing Lines of Equations 107
Equation of Parallel or Perpendicular Lines 108
Equations of Horizontal and Vertical Lines 109
Graphing Absolute Value Equations 110
Answers of Worksheets – Chapter 8 111

Chapter 9: Monomials and polynomials 117
GCF of Monomials 118
Factoring Quadratics 119
Factoring by Grouping 120
GCF and Powers of Monomials 121
Writing Polynomials in Standard Form 122
Simplifying Polynomials 123
Adding and Subtracting Polynomials 124
Multiplying a Polynomial and a Monomial 125
Multiplying Binomials 126
Factoring Trinomials 127

Operations with Polynomials 128
Answers of Worksheets – Chapter 9 129

Chapter 10: Functions Operations and Quadratic 135
Evaluating Function 136
Adding and Subtracting Functions 137
Multiplying and Dividing Functions 138
Composition of Functions 139
Quadratic Equation 140
Solving Quadratic Equations 141
Quadratic Formula and the Discriminant 142
Quadratic Inequalities 143
Graphing Quadratic Functions 144
Domain and Range of Radical Functions 145
Solving Radical Equations 146
Answers of Worksheets – Chapter 10 147

Chapter 11: Complex Numbers 153
Adding and Subtracting Complex Numbers 154
Multiplying and Dividing Complex Numbers 155
Graphing Complex Numbers 156
Rationalizing Imaginary Denominators 157
Answers of Worksheets – Chapter 11 158

Chapter 12: Logarithms 159
Rewriting Logarithms 160
Evaluating Logarithms 161
Properties of Logarithms 162
Natural Logarithms 163
Exponential Equations and Logarithms 164
Solving Logarithmic Equations 165
Answers of Worksheets – Chapter 12 166

Chapter 13: Geometry and Solid Figures 169
Angles 170
Pythagorean Relationship 171
Triangles 172

Polygons 173

Trapezoids 174

Circles 175

Cubes 176

Rectangular Prism 177

Cylinder 178

Pyramids and Cone 179

Answers of Worksheets – Chapter 13 180

Chapter 14: Statistics and Probability **183**

Mean and Median 184

Mode and Range 185

Probability Problems 186

Factorials 187

Combinations and Permutations 188

Answers of Worksheets – Chapter 14 189

CLEP College Mathematics Test Review **191**

CLEP College Mathematics Test Answer Sheets 193

CLEP College Mathematics Practice Test 1 195

CLEP College Mathematics Practice Test 2 213

Answers and Explanations **229**

Answer Key 229

Practice Tests 1 231

Practice Tests 2 243

Chapter 1:
Integers and Number Theory

Topics that you will practice in this chapter:

- ✓ Adding and Subtracting Integers
- ✓ Multiplying and Dividing Integers
- ✓ Order of Operations
- ✓ Ordering Integers and Numbers
- ✓ Integers and Absolute Value
- ✓ Factoring Numbers
- ✓ Greatest Common Factor (GCF)
- ✓ Least Common Multiple (LCM)
- ✓ Sets

"Wherever there is number, there is beauty." –Proclus

Adding and Subtracting Integers

✍ **Find each sum.**

1) $15 + (-35) =$

2) $(-28) + (-29) =$

3) $19 + (-27) =$

4) $57 + (-64) =$

5) $(-14) + (-19) + 64 =$

6) $54 + (-36) + 19 =$

7) $46 + (-30) + (-33) + 29 =$

8) $(-40) + (-70) + 28 + 55 =$

9) $60 + (-65) + (83 - 72) =$

10) $49 + (-55) + (90 - 67) =$

✍ **Find each difference.**

11) $(-32) - (-7) =$

12) $40 - (-12) =$

13) $(-60) - 56 =$

14) $27 - (-17) =$

15) $58 - (76 - 29) =$

16) $19 - (-14) - (-22) =$

17) $(39 + 15) - (-46) =$

18) $49 - 17 - (-13) =$

19) $85 - 45 - (-18) =$

20) $78 - (-35) - (-63) =$

21) $89 - (-11) - (-26) =$

22) $(19 - 50) - (-95) =$

23) $46 - 49 - (-87) =$

24) $120 - (98 + 24) - (-38) =$

25) $112 - (-102) + (-81) =$

26) $108 - (-42) + (-89) =$

Multiplying and Dividing Integers

✎ **Find each product.**

1) $(-7) \times (-9) =$

2) $(-5) \times 6 =$

3) $10 \times (-15) =$

4) $(-9) \times (-25) =$

5) $(-7) \times (-12) \times 13 =$

6) $(15 - 4) \times (-11) =$

7) $25 \times (-4) \times (-5) =$

8) $(85 + 10) \times (-11) =$

9) $12 \times (-19 + 12) \times 5 =$

10) $(-15) \times (-18) \times (-20) =$

✎ **Find each quotient.**

11) $85 \div (-5) =$

12) $(-90) \div (-15) =$

13) $(-121) \div (-11) =$

14) $99 \div (-33) =$

15) $(-114) \div 2 =$

16) $(-208) \div (-16) =$

17) $198 \div (-11) =$

18) $(-364) \div (-14) =$

19) $255 \div (-15) =$

20) $(-378) \div (18) =$

21) $(-184) \div (-8) =$

22) $-437 \div (-23) =$

23) $(-570) \div (-19) =$

24) $480 \div (-32) =$

25) $(-546) \div (-21) =$

26) $(486) \div (-54) =$

Order of Operations

✏️ **Evaluate each expression.**

1) $7 + (5 \times 8) =$

2) $16 - (6 \times 9) =$

3) $(17 \times 5) + 12 =$

4) $(24 - 12) - (11 \times 4) =$

5) $35 + (18 \div 3) =$

6) $(27 \times 3) \div 3 =$

7) $(88 \div 4) \times (-5) =$

8) $(9 \times 9) + (86 - 52) =$

9) $78 + (5 \times 12) + 14 =$

10) $(60 \times 4) \div (4 + 2) =$

11) $(-15) + (14 \times 4) + 18 =$

12) $(14 \times 5) - (56 \div 7) =$

13) $(7 \times 9 \div 3) - (32 + 21) =$

14) $(45 + 11 - 14) \times 2 - 15 =$

15) $(40 - 18 + 20) \times (75 \div 3) =$

16) $75 + (54 - (45 \div 9)) =$

17) $(12 + 15 - 24) + (44 \div 4) =$

18) $(78 - 19) + (27 - 10 + 7) =$

19) $(18 \times 3) + (17 \times 9) - 52 =$

20) $65 + 17 - (45 \times 2) + 40 =$

Ordering Integers and Numbers

✎ **Order each set of integers from least to greatest.**

1) $17, -15, -8, 0, 9$ ___, ___, ___, ___, ___, ___

2) $-14, -26, 17, 42, 39$ ___, ___, ___, ___, ___, ___

3) $32, -15, -69, 41, -80$ ___, ___, ___, ___, ___, ___

4) $-49, -65, 35, -21, 68$ ___, ___, ___, ___, ___, ___

5) $69, -32, 10, -45, 24$ ___, ___, ___, ___, ___, ___

6) $108, 76, -59, 87, -78$ ___, ___, ___, ___, ___, ___

✎ **Order each set of integers from greatest to least.**

7) $62, 98, -7, -19, -1$ ___, ___, ___, ___, ___, ___

8) $34, 35, -24, -46, 56$ ___, ___, ___, ___, ___, ___

9) $35, -96, -58, 17, -34$ ___, ___, ___, ___, ___, ___

10) $37, 12, -26, -13, 52$ ___, ___, ___, ___, ___, ___

11) $-12, 66, -18, -28, 54$ ___, ___, ___, ___, ___, ___

12) $-100, -85, -30, 5, 9$ ___, ___, ___, ___, ___, ___

Integers and Absolute Value

✎ **Write absolute value of each number.**

1) $|-19| =$

2) $|-32| =$

3) $|-50| =$

4) $|31| =$

5) $|57| =$

6) $|-76| =$

7) $|42| =$

8) $|101| =$

9) $|28| =$

10) $|-49| =$

11) $|-13|$

12) $|78| =$

13) $|100| =$

14) $|0| =$

15) $|-105| =$

16) $|-77| =$

17) $88 =$

18) $|-29| =$

19) $|112| =$

20) $|-120| =$

✎ **Evaluate the value.**

21) $|-5| - \frac{|-40|}{8} =$

22) $18 - |4 - 19| - |-15| =$

23) $\frac{|-72|}{9} \times |-9| =$

24) $\frac{|6 \times (-8)|}{3} \times \frac{|-21|}{7} =$

25) $|5 \times (-9)| + \frac{|-110|}{11} =$

26) $\frac{|-96|}{12} \times \frac{|-27|}{9} =$

27) $|-19 + 12| \times \frac{|-12 \times 13|}{7}$

28) $\frac{|-19 \times 6|}{3} \times |-11| =$

Factoring Numbers

✎ **List all positive factors of each number.**

1) 6

2) 21

3) 28

4) 26

5) 46

6) 45

7) 48

8) 50

9) 52

10) 63

11) 70

12) 72

13) 78

14) 80

15) 82

16) 88

17) 90

18) 93

19) 95

20) 96

21) 98

22) 102

23) 124

24) 125

Greatest Common Factor

Find the GCF for each number pair.

1) 6, 2

2) 8, 4

3) 5, 3

4) 6, 4

5) 7, 5

6) 8, 18

7) 14, 21

8) 6, 14

9) 9, 15

10) 4, 18

11) 14, 18

12) 25, 30

13) 27, 45

14) 36, 18

15) 9, 12

16) 11, 8

17) 28, 21

18) 56, 72

19) 34, 51

20) 6, 18, 27

21) 2, 9, 8

22) 10, 12, 24

23) 5, 14, 21

24) 72, 9, 18

Least Common Multiple

✎ Find the LCM for each number pair.

1) 6, 5

2) 8, 18

3) 9, 15

4) 15, 20

5) 20, 25

6) 22, 33

7) 6, 28

8) 8, 14

9) 21, 28

10) 14, 28

11) 9, 30

12) 7, 12

13) 12, 36

14) 9, 54

15) 42, 21

16) 40, 16

17) 12, 42

18) 13, 11

19) 32, 72

20) 15, 27

21) 24, 44

22) 8, 12, 42

23) 2, 6, 11

24) 15, 25, 30

Sets

✏ Given A = {1, 2, 3, 8, 12}, B = {2, 4, 5, 7}, and C = {5, 7, 9, 11}, find:

1) A ∪ B _____

2) A ∪ C _____

3) B ∪ C _____

4) A ∩ B _____

5) A ∩ C _____

6) B ∩ C _____

7) (A ∪ B) ∪ C _____

8) (A ∪ B) ∩ C _____

9) (A ∩ B) ∩ C _____

10) (B ∪ C) ∩ A _____

✏ Refer to the diagram below to find each set.

11) A ∪ B _____

12) A ∪ C _____

13) B ∪ C _____

14) A ∩ B _____

15) A ∩ C _____

16) B ∩ C _____

17) (A ∪ B) ∪ C _____

18) (A ∪ B) ∩ C _____

19) (A ∩ B) ∩ C _____

20) (B ∪ C) ∩ A _____

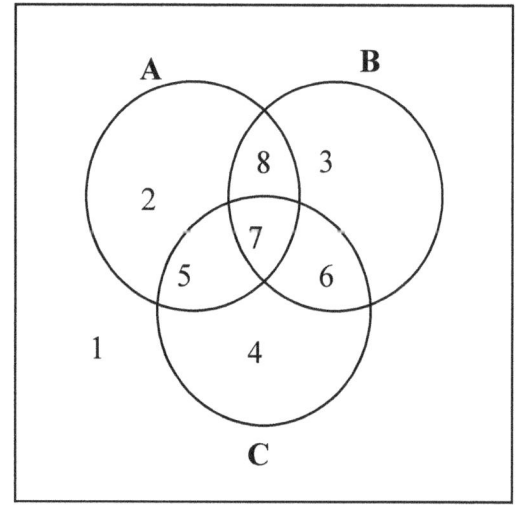

Answers of Worksheets – Chapter 1

Adding and Subtracting Integers

1) −20
2) −57
3) −8
4) −7
5) 31
6) 37
7) 12
8) −27
9) 6
10) 17
11) −25
12) 52
13) −116
14) 44
15) 11
16) 55
17) 100
18) 45
19) 58
20) 176
21) 126
22) 64
23) 84
24) 36
25) 133
26) 61

Multiplying and Dividing Integers

1) 63
2) −30
3) −150
4) 225
5) 1,092
6) −121
7) 500
8) −1,045
9) −420
10) −5,400
11) −17
12) 6
13) 11
14) −3
15) −57
16) 13
17) −18
18) 26
19) −17
20) −21
21) 23
22) 19
23) 30
24) −15
25) 26
26) −9

Order of Operations

1) 47
2) −38
3) 97
4) −32
5) 41
6) 27
7) −110
8) 115
9) 152
10) 40
11) 59
12) 62
13) −32
14) 69
15) 1,050
16) 124
17) 14
18) 83
19) 155
20) 32

Ordering Integers and Numbers

1) −15, −8, 0, 9, 17
2) −26, −14, 17, 39, 42
3) −80, −69, −15, 32, 41
4) −65, −49, −21, 35, 68
5) −45, −32, 10, 24, 69
6) −78, −59, 76, 87, 108
7) 98, 62, −1, −7, −19
8) 56, 35, 34, −24, −46
9) 35, 17, −34, −58, −96
10) 52, 37, 12, −13, −26
11) 66, 54, −12, −18, −28
12) 9, 5, −30, −85, −100

Integers and Absolute Value

1) 19
2) 32
3) 50
4) 31
5) 57
6) 76
7) 42
8) 101
9) 28
10) 49
11) 13
12) 78
13) 100
14) 0
15) 105
16) 77
17) 88
18) 29
19) 112
20) 120
21) 0
22) -12
23) 72
24) 48
25) 55
26) 24
27) 156
28) 418

Factoring Numbers

1) 1, 2, 3, 6
2) 1, 3, 7, 21
3) 1, 2, 4, 7, 14, 28
4) 1, 2, 13, 26
5) 1, 2, 23, 46
6) 1, 3, 5, 9, 15, 45
7) 1, 2, 3, 4, 6, 8, 12, 16, 24, 48
8) 1, 2, 5, 10, 25, 50
9) 1, 2, 4, 5, 13, 26, 52
10) 1, 3, 7, 9, 21, 63
11) 1, 2, 5, 7, 10, 14, 35, 70
12) 1, 2, 3, 4, 6, 8, 9, 12, 18, 24, 36, 72
13) 1, 2, 3, 6, 13, 26, 39, 78
14) 1, 2, 4, 5, 8, 10, 16, 20, 40, 80
15) 1, 2, 41, 82
16) 1, 2, 4, 8, 11, 22, 44, 88
17) 1, 2, 3, 5, 6, 9, 10, 15, 18, 30, 45, 90
18) 1, 3, 31, 93
19) 1, 5, 19, 95
20) 1, 2, 3, 4, 6, 8, 12, 16, 24, 32, 48, 96
21) 1, 2, 7, 14, 49, 98
22) 1, 2, 3, 6, 17, 34, 51, 102
23) 1, 2, 4, 31, 62, 124
24) 1, 5, 25, 125

Greatest Common Factor

1) 2
2) 4
3) 1
4) 2
5) 1
6) 2
7) 7
8) 2
9) 3
10) 2
11) 2
12) 5
13) 9
14) 18
15) 3
16) 1
17) 7
18) 8
19) 17
20) 3
21) 1
22) 2
23) 1
24) 9

Least Common Multiple

1) 30
2) 72
3) 45
4) 60
5) 100
6) 66
7) 84
8) 56
9) 84
10) 28
11) 90
12) 84
13) 36
14) 54
15) 42
16) 80
17) 84
18) 143
19) 288
20) 135
21) 264
22) 168
23) 66
24) 150

Sets

1) {1, 2, 3, 4, 5, 7, 8, 12}
2) {1, 2, 3, 5, 7, 8, 9, 11, 12}
3) {2, 4, 5, 7, 9, 11}
4) {2}
5) { } (empty set)
6) {5, 7}
7) {1, 2, 3, 4, 5, 7, 8, 9, 11, 12}
8) {5, 7}
9) { } (empty set)
10) {2}
11) {2, 3, 5, 6, 7, 8}
12) {2, 4, 5, 6, 7, 8}
13) {3, 4, 5, 6, 7, 8}
14) {7, 8}
15) {5, 7}
16) {6, 7}
17) {2, 3, 4, 5, 6, 7, 8}
18) {5, 6, 7}
19) {7}
20) {5, 7, 8}

Chapter 2: Fractions and Decimals

Topics that you will practice in this chapter:

- ✓ Simplifying Fractions
- ✓ Adding and Subtracting Fractions
- ✓ Multiplying and Dividing Fractions
- ✓ Adding and Subtract Mixed Numbers
- ✓ Multiplying and Dividing Mixed Numbers
- ✓ Adding and Subtracting Decimals
- ✓ Multiplying and Dividing Decimals
- ✓ Comparing Decimals
- ✓ Rounding Decimals

"A Man is like a fraction whose numerator is what he is and whose denominator is what he thinks of himself. The larger the denominator, the smaller the fraction." –Tolstoy

Simplifying Fractions

✎ **Simplify each fraction to its lowest terms.**

1) $\frac{8}{16} =$

2) $\frac{28}{35} =$

3) $\frac{27}{36} =$

4) $\frac{70}{140} =$

5) $\frac{13}{52} =$

6) $\frac{38}{57} =$

7) $\frac{64}{80} =$

8) $\frac{21}{84} =$

9) $\frac{85}{170} =$

10) $\frac{120}{168} =$

11) $\frac{31}{124} =$

12) $\frac{48}{96} =$

13) $\frac{98}{112} =$

14) $\frac{99}{110} =$

15) $\frac{51}{153} =$

16) $\frac{40}{112} =$

17) $\frac{90}{225} =$

18) $\frac{44}{297} =$

19) $\frac{54}{279} =$

20) $\frac{320}{720} =$

21) $\frac{70}{560} =$

✎ **Find the answer for each problem.**

22) Which of the following fractions equal to $\frac{3}{7}$? ____

   A. $\frac{24}{63}$ B. $\frac{51}{109}$ C. $\frac{51}{119}$ D. $\frac{240}{630}$

23) Which of the following fractions equal to $\frac{7}{8}$? ____

   A. $\frac{182}{208}$ B. $\frac{175}{208}$ C. $\frac{182}{216}$ D. $\frac{49}{64}$

24) Which of the following fractions equal to $\frac{2}{9}$? ____

   A. $\frac{64}{126}$ B. $\frac{46}{207}$ C. $\frac{48}{207}$ D. $\frac{56}{208}$

Adding and Subtracting Fractions

Find the sum.

1) $\frac{5x}{8} + \frac{3x}{8} =$

2) $\frac{x}{2} + \frac{x}{7} =$

3) $\frac{y}{3} + \frac{y}{4} =$

4) $\frac{3x}{8} + \frac{2x}{5} =$

5) $\frac{xy}{5} + \frac{2xy}{7} =$

6) $\frac{2x}{9} + \frac{4x}{9} =$

7) $\frac{a}{4} + \frac{2a}{3} =$

8) $\frac{2}{x} + \frac{4}{x} =$

9) $\frac{1}{a} + \frac{2}{b} =$

10) $\frac{3b}{5} + \frac{2b}{7} =$

11) $\frac{a}{y} + \frac{3a}{y} =$

12) $\frac{3}{x} + \frac{1}{2x} =$

Find the difference.

13) $\frac{x}{3} - \frac{x}{6} =$

14) $\frac{2x}{5} - \frac{3x}{8} =$

15) $\frac{x}{7} - \frac{y}{7} =$

16) $\frac{2x}{7} - \frac{x}{6} =$

17) $\frac{5a}{9} - \frac{2a}{5} =$

18) $\frac{2ab}{3} - \frac{ab}{6} =$

19) $\frac{1}{x} - \frac{1}{3x} =$

20) $\frac{4}{y} - \frac{3}{4y} =$

21) $\frac{5}{x} - \frac{2y}{xy} =$

22) $\frac{8}{ab} - \frac{5}{3ab} =$

23) $\frac{2a}{y} - \frac{a}{3y} =$

24) $\frac{5}{b} - \frac{2}{3b} =$

25) $\frac{2a}{b} - \frac{a}{b} =$

26) $\frac{3}{a} - \frac{2}{b} =$

27) $\frac{4a}{b} - \frac{2a}{3b} =$

28) $\frac{6}{xy} - \frac{7}{2xy} =$

29) $\frac{2}{a} - \frac{1}{4a} =$

30) $\frac{2a}{3b} - \frac{4a}{9b} =$

Multiplying and Dividing Fractions

✏️ Find the value of each expression in lowest terms.

1) $\dfrac{3}{a} \times \dfrac{5}{3} =$

2) $\dfrac{2}{3b} \times \dfrac{9}{2} =$

3) $\dfrac{a}{15} \times \dfrac{5}{2a} =$

4) $\dfrac{x}{3a} \times \dfrac{9a}{6x} =$

5) $\dfrac{x}{12} \times \dfrac{6}{y} =$

6) $\dfrac{7}{x} \times \dfrac{x}{14} =$

7) $\dfrac{10}{3a} \times \dfrac{6}{20} =$

8) $\dfrac{4a}{b} \times \dfrac{2b}{5} =$

9) $\dfrac{2ab}{7} \times \dfrac{14}{6ab} =$

10) $\dfrac{4a}{5b} \times \dfrac{15}{2a} =$

11) $\dfrac{ab}{21} \times \dfrac{7}{a} =$

12) $\dfrac{a}{cd} \times \dfrac{2bc}{a} =$

✏️ Find the value of each expression in lowest terms.

13) $\dfrac{a}{2} \div \dfrac{a}{4} =$

14) $\dfrac{b}{3} \div \dfrac{b}{9} =$

15) $\dfrac{a}{b} \div \dfrac{3}{b} =$

16) $\dfrac{2a}{15} \div \dfrac{4a}{5} =$

17) $\dfrac{1}{a} \div \dfrac{b}{3a} =$

18) $\dfrac{4a}{3b} \div \dfrac{a}{2b} =$

19) $\dfrac{a}{8} \div \dfrac{3a}{16} =$

20) $\dfrac{3b}{20} \div \dfrac{6b}{15a} =$

21) $\dfrac{x}{12y} \div \dfrac{2x}{9y} =$

22) $\dfrac{25}{x} \div \dfrac{50}{2x} =$

23) $\dfrac{16}{5ab} \div \dfrac{32}{ab} =$

24) $\dfrac{7a}{b} \div \dfrac{8a}{b} =$

25) $\dfrac{5}{x} \div \dfrac{3y}{x} =$

26) $\dfrac{2a}{21} \div \dfrac{a}{14} =$

27) $\dfrac{ab}{x} \div \dfrac{a}{x} =$

28) $\dfrac{6}{a} \div \dfrac{3b}{2a} =$

29) $\dfrac{9}{16a} \div \dfrac{3}{8ab} =$

30) $\dfrac{24}{xy} \div \dfrac{12}{y} =$

Adding and Subtracting Mixed Numbers

🖎 **Find the sum.**

1) $3\frac{5}{6} + 2\frac{1}{3} =$

2) $4\frac{2}{5} + 1\frac{1}{5} =$

3) $5\frac{1}{8} + 6\frac{3}{4} =$

4) $2\frac{2}{3} + 3\frac{1}{2} =$

5) $3\frac{4}{5} + 3\frac{2}{15} =$

6) $8\frac{1}{16} + 3\frac{3}{8} =$

7) $4\frac{3}{5} + 4\frac{1}{6} =$

8) $7\frac{3}{4} + 3\frac{5}{6} =$

9) $8\frac{5}{6} + 2\frac{2}{7} =$

10) $11\frac{3}{16} + 3\frac{5}{24} =$

🖎 **Find the difference.**

11) $3\frac{3}{4} - 2\frac{1}{4} =$

12) $5\frac{1}{7} - 3\frac{1}{7} =$

13) $4\frac{1}{3} - 1\frac{1}{9} =$

14) $7\frac{1}{6} - 3\frac{1}{12} =$

15) $6\frac{1}{3} - 2\frac{5}{18} =$

16) $8\frac{1}{4} - 5\frac{1}{8} =$

17) $9\frac{1}{2} - 6\frac{1}{5} =$

18) $11\frac{7}{15} - 8\frac{1}{30} =$

19) $12\frac{3}{5} - 7\frac{2}{7} =$

20) $18\frac{1}{8} - 14\frac{3}{16} =$

21) $12\frac{2}{3} - 11\frac{7}{15} =$

22) $3\frac{1}{5} - 1\frac{1}{2} =$

23) $14\frac{3}{5} - 6\frac{4}{5} =$

24) $17\frac{1}{4} - 14\frac{8}{9} =$

25) $24\frac{3}{9} - 15\frac{1}{18} =$

26) $28\frac{3}{7} - 19\frac{5}{6} =$

Multiplying and Dividing Mixed Numbers

🖎 **Find the product.**

1) $2\frac{1}{3} \times 4\frac{1}{2} =$

2) $4\frac{1}{5} \times 2\frac{1}{3} =$

3) $7\frac{2}{3} \times 3\frac{3}{5} =$

4) $9\frac{2}{7} \times 3\frac{1}{8} =$

5) $5\frac{4}{11} \times 4\frac{1}{3} =$

6) $7\frac{3}{8} \times 5\frac{4}{9} =$

7) $9\frac{2}{3} \times 11\frac{5}{6} =$

8) $8\frac{3}{5} \times 7\frac{4}{9} =$

9) $5\frac{1}{9} \times 9\frac{5}{8} =$

10) $10\frac{2}{7} \times 2\frac{5}{8} =$

🖎 **Find the quotient.**

11) $2\frac{1}{8} \div 1\frac{3}{8} =$

12) $4\frac{1}{6} \div 2\frac{1}{3} =$

13) $7\frac{1}{3} \div 3\frac{3}{4} =$

14) $4\frac{5}{8} \div 1\frac{1}{2} =$

15) $6\frac{5}{12} \div 4\frac{1}{6} =$

16) $5\frac{7}{18} \div 5\frac{1}{6} =$

17) $6\frac{5}{21} \div 2\frac{3}{7} =$

18) $8\frac{1}{7} \div 8\frac{1}{14} =$

19) $10\frac{1}{4} \div 3\frac{2}{5} =$

20) $15\frac{1}{3} \div 5\frac{2}{9} =$

21) $12\frac{1}{3} \div 6\frac{1}{2} =$

22) $18\frac{1}{9} \div 18\frac{1}{6} =$

23) $10\frac{3}{4} \div 5\frac{2}{5} =$

24) $11\frac{1}{3} \div 8\frac{4}{5} =$

25) $9\frac{1}{6} \div 3\frac{2}{7} =$

26) $7\frac{1}{3} \div 3\frac{7}{11} =$

Adding and Subtracting Decimals

✍ **Add and subtract decimals.**

1) 52.18
   − 21.27
   _____

2) 49.34
   + 25.24
   _____

3) 48.60
   + 35.75
   _____

4) 65.84
   − 35.49
   _____

5) 54.57
   + 18.37
   _____

6) 90.45
   − 28.75
   _____

7) 98.12
   − 45.55
   _____

8) 48.99
   + 57.67
   _____

9) 158.05
   − 78.98
   _____

✍ **Find the missing number.**

10) ___ + 4.9 = 6.5

11) 5.15 + ___ = 6.43

12) 8.09 + ___ = 11.84

13) 8.88 − ___ = 6.78

14) ___ − 1.59 = 3.71

15) ___ − 19.98 = 8.17

16) 38.89 + ___ = 41.32

17) ___ − 35.99 = 1.80

18) ___ + 39.08 = 41.36

19) 98.98 + ___ = 123.68

Multiplying and Dividing Decimals

✎ **Find the product.**

1) 0.6 × 0.8 =

2) 2.5 × 0.9 =

3) 0.87 × 0.4 =

4) 0.15 × 0.75 =

5) 0.95 × 0.7 =

6) 1.57 × 0.9 =

7) 5.85 × 1.3 =

8) 12.5 × 4.5 =

9) 19.8 × 7.32 =

10) 85.1 × 1.5 =

11) 79.5 × 11.2 =

12) 86.9 × 21.5 =

✎ **Find the quotient.**

13) 3.25 ÷ 10 =

14) 24.5 ÷ 100 =

15) 3.9 ÷ 3 =

16) 91.2 ÷ 0.6 =

17) 29.2 ÷ 0.4 =

18) 38.7 ÷ `9 =

19) 297.8 ÷ 1,000 =

20) 53.55 ÷ 0.7 =

21) 345.45 ÷ 0.1 =

22) 70.27 ÷ 0.25 =

23) 28.968 ÷ 0.3 =

24) 86.34 ÷ 0.06 =

Comparing Decimals

✎ **Write the correct comparison symbol (>, < or =).**

1) 0.80 ☐ 0.080

2) 0.086 ☐ 0.86

3) 7.090 ☐ 7.09

4) 3.25 ☐ 3.06

5) 4.09 ☐ 0.490

6) 6.06 ☐ 6.6

7) 6.08 ☐ 6.080

8) 4.05 ☐ 4.2

9) 12.35 ☐ 12.198

10) 0.957 ☐ 0.0957

11) 25.24 ☐ 25.240

12) 0.742 ☐ 0.752

13) 14.09 ☐ 14.10

14) 17.45 ☐ 17.154

15) 11.44 ☐ 11.439

16) 15.41 ☐ 15.410

17) 21.43 ☐ 21.043

18) 8.098 ☐ 8.90

19) 16.044 ☐ 16.040

20) 32.35 ☐ 32.350

Rounding Decimals

✎ **Round each decimal to the nearest whole number.**

1) 56.27   3) 18.32   5) 7.90

2) 5.9   4) 4.8   6) 57.7

✎ **Round each decimal to the nearest tenth.**

7) 42.785   9) 96.586   11) 27.198

8) 15.224   10) 101.78   12) 96.87

✎ **Round each decimal to the nearest hundredth.**

13) 9.648   15) 89.2882   17) 68.229

14) 27.819   16) 120.912   18) 85.642

✎ **Round each decimal to the nearest thousandth.**

19) 19.88486   21) 145.9322   23) 189.0991

20) 46.72611   22) 210.1581   24) 121.76798

Answers of Worksheets – Chapter 2

Simplifying Fractions

1) $\frac{1}{2}$
2) $\frac{4}{5}$
3) $\frac{3}{4}$
4) $\frac{1}{2}$
5) $\frac{1}{4}$
6) $\frac{2}{3}$
7) $\frac{4}{5}$
8) $\frac{1}{4}$
9) $\frac{1}{2}$
10) $\frac{5}{7}$
11) $\frac{1}{4}$
12) $\frac{1}{2}$
13) $\frac{7}{8}$
14) $\frac{9}{10}$
15) $\frac{1}{3}$
16) $\frac{5}{14}$
17) $\frac{2}{5}$
18) $\frac{4}{27}$
19) $\frac{6}{31}$
20) $\frac{4}{9}$
21) $\frac{1}{8}$
22) C
23) A
24) B

Adding and Subtracting Fractions

1) $\frac{8x}{8} = x$
2) $\frac{9x}{14}$
3) $\frac{7x}{12}$
4) $\frac{31x}{40}$
5) $\frac{17xy}{35}$
6) $\frac{2x}{3}$
7) $\frac{11a}{12}$
8) $\frac{6}{x}$
9) $\frac{a+2b}{ab}$
10) $\frac{31b}{35}$
11) $\frac{4a}{y}$
12) $\frac{7}{2x}$
13) $\frac{x}{6}$
14) $\frac{x}{40}$
15) $\frac{x-y}{7}$
16) $\frac{5x}{42}$
17) $\frac{7a}{45}$
18) $\frac{ab}{2}$
19) $\frac{2}{3x}$
20) $\frac{13}{4y}$
21) $\frac{3}{x}$
22) $\frac{19}{3ab}$
23) $\frac{5a}{3y}$
24) $\frac{13}{3b}$
25) $\frac{a}{b}$
26) $\frac{3b-2a}{ab}$
27) $\frac{10a}{3b}$
28) $\frac{5}{2xy}$
29) $\frac{7}{4a}$
30) $\frac{2a}{9b}$

Multiplying and Dividing Fractions

1) $\frac{5}{a}$
2) $\frac{3}{b}$
3) $\frac{1}{6}$
4) $\frac{1}{2}$
5) $\frac{x}{2y}$
6) $\frac{1}{2}$
7) $\frac{1}{a}$
8) $\frac{8a}{5}$
9) $\frac{2}{3}$

CLEP College Math Workbook

10) $\dfrac{6}{b}$

11) $\dfrac{b}{3}$

12) $\dfrac{2b}{d}$

13) 2

14) 3

15) $\dfrac{a}{3}$

16) $\dfrac{1}{6}$

17) $\dfrac{3}{b}$

18) $\dfrac{8}{3}$

19) $\dfrac{2}{3}$

20) $\dfrac{3a}{8}$

21) $\dfrac{3}{8}$

22) 1

23) $\dfrac{1}{10}$

24) $\dfrac{7}{8}$

25) $\dfrac{5}{3y}$

26) $\dfrac{4}{3}$

27) b

28) $\dfrac{4}{b}$

29) $\dfrac{3b}{2}$

30) $\dfrac{2}{x}$

Adding and Subtracting Mixed Numbers

1) $6\dfrac{1}{6}$

2) $5\dfrac{3}{5}$

3) $11\dfrac{7}{8}$

4) $6\dfrac{1}{6}$

5) $6\dfrac{14}{15}$

6) $11\dfrac{7}{16}$

7) $8\dfrac{23}{30}$

8) $11\dfrac{7}{12}$

9) $11\dfrac{5}{42}$

10) $14\dfrac{19}{48}$

11) $1\dfrac{1}{2}$

12) 2

13) $3\dfrac{2}{9}$

14) $4\dfrac{1}{12}$

15) $4\dfrac{1}{18}$

16) $3\dfrac{1}{8}$

17) $3\dfrac{3}{10}$

18) $3\dfrac{13}{30}$

19) $5\dfrac{11}{35}$

20) $3\dfrac{15}{16}$

21) $1\dfrac{1}{5}$

22) $1\dfrac{7}{10}$

23) $7\dfrac{4}{5}$

24) $2\dfrac{13}{36}$

25) $9\dfrac{5}{18}$

26) $8\dfrac{25}{42}$

Multiplying and Dividing Mixed Numbers

1) $10\dfrac{1}{2}$

2) $9\dfrac{4}{5}$

3) $27\dfrac{3}{5}$

4) $29\dfrac{1}{56}$

5) $23\dfrac{8}{33}$

6) $40\dfrac{11}{72}$

7) $144\dfrac{7}{18}$

8) $64\dfrac{1}{45}$

9) $49\dfrac{7}{36}$

10) 27

11) $1\dfrac{6}{11}$

12) $1\dfrac{11}{14}$

13) $1\dfrac{43}{45}$

14) $3\dfrac{1}{12}$

15) $1\dfrac{27}{50}$

16) $1\dfrac{4}{93}$

17) $2\dfrac{29}{51}$

18) $1\dfrac{1}{113}$

19) $3\dfrac{1}{68}$

20) $2\dfrac{44}{47}$

21) $1\dfrac{35}{39}$

22) $\frac{326}{327}$

23) $1\frac{107}{108}$

24) $1\frac{19}{66}$

25) $2\frac{109}{138}$

26) $2\frac{1}{60}$

Adding and Subtracting Decimals

1) 30.91
2) 74.58
3) 84.35
4) 30.35
5) 72.94
6) 61.7
7) 52.57
8) 106.66
9) 79.07
10) 1.6
11) 1.28
12) 3.75
13) 2.1
14) 5.3
15) 28.15
16) 2.43
17) 37.79
18) 2.28
19) 24.7

Multiplying and Dividing Decimals

1) 0.48
2) 2.25
3) 0.348
4) 0.1125
5) 0.665
6) 1.413
7) 7.605
8) 56.25
9) 144.936
10) 127.65
11) 890.4
12) 1,868.35
13) 0.325
14) 0.245
15) 1.3
16) 152
17) 73
18) 4.3
19) 0.2978
20) 76.5
21) 3,454.5
22) 281.08
23) 96.56
24) 1,439

Comparing Decimals

1) >
2) <
3) =
4) >
5) >
6) <
7) =
8) <
9) >
10) >
11) =
12) <
13) <
14) >
15) >
16) =
17) >
18) <
19) >
20) =

Rounding Decimals

1) 56
2) 6
3) 18
4) 5
5) 8
6) 58
7) 42.8
8) 15.2
9) 96.6
10) 101.8
11) 27.2
12) 96.9
13) 9.65
14) 27.82
15) 89.29
16) 120.91
17) 68.23
18) 85.64
19) 19.885
20) 46.726
21) 145.932
22) 210.158
23) 189.099
24) 121.768

Chapter 3: Proportions, Ratios, and Percent

Topics that you will practice in this chapter:

- ✓ Simplifying Ratios
- ✓ Proportional Ratios
- ✓ Similarity and Ratios
- ✓ Ratio and Rates Word Problems
- ✓ Percentage Calculations
- ✓ Percent Problems
- ✓ Discount, Tax and Tip
- ✓ Percent of Change
- ✓ Simple Interest

Without mathematics, there's nothing you can do. Everything around you is mathematics. Everything around you is numbers." – Shakuntala Devi

Simplifying Ratios

✏ **Reduce each ratio.**

1) $15:20 = \_\_\_:\_\_\_$

2) $9:90 = \_\_\_:\_\_\_$

3) $24:42 = \_\_\_:\_\_\_$

4) $7:21 = \_\_\_:\_\_\_$

5) $11:110 \_\_\_:\_\_\_$

6) $8:64 = \_\_\_:\_\_\_$

7) $18:72 = \_\_\_:\_\_\_$

8) $10:25 = \_\_\_:\_\_\_$

9) $7:42 = \_\_\_:\_\_\_$

10) $49:63 = \_\_\_:\_\_\_$

11) $12:18 \_\_\_:\_\_\_$

12) $35:10 \_\_\_:\_\_\_$

13) $150:15 \_\_\_:\_\_\_$

14) $2.4:3.2 \_\_\_:\_\_\_$

15) $7:56 = \_\_\_:\_\_\_$

16) $45:63 \_\_\_:\_\_\_$

17) $77:99 \_\_\_:\_\_\_$

18) $39:13 \_\_\_:\_\_\_$

19) $15:45 \_\_\_:\_\_\_$

20) $84:12 \_\_\_:\_\_\_$

21) $25:5 \_\_\_:\_\_\_$

22) $70:56 \_\_\_:\_\_\_$

23) $70:140 \_\_\_:\_\_\_$

24) $1.2:36 \_\_\_:\_\_\_$

✏ **Write each ratio as a fraction in simplest form.**

25) $7:14 =$

26) $27:45 =$

27) $24:56 =$

28) $16:48 =$

29) $22:66 =$

30) $21:98 =$

31) $34:68 =$

32) $6:30 =$

33) $35:84 =$

34) $12:54 =$

35) $88:104 =$

36) $36:81 =$

37) $1.5:18 =$

38) $4.5:16.5 =$

39) $5:75 =$

40) $3.1:12.4 =$

41) $1.6:6.4 =$

42) $0.25:1.25 =$

43) $8.8:16.4 =$

44) $0.75:6.75 =$

45) $1.8:3 =$

Proportional Ratios

🍃 **Fill in the blanks; Calculate each proportion.**

1) $3:8 = \_\_ : 32$

2) $1:2 = 45 : \_\_$

3) $1:11 = \_\_ : 55$

4) $9:12 = 18 : \_\_$

5) $9:7 = 81 : \_\_$

6) $2:8 = \_\_ : 56$

7) $2.3:1.2 = \_\_ : 12$

8) $0.5:2 = \_\_ : 32$

9) $1.6:2 = \_\_ : 60$

10) $2.5:4.5 = \_\_ : 90$

11) $3.8:7.1 = 7.6 : \_\_$

12) $5.5:6 = 16.5 : \_\_$

🍃 **State if each pair of ratios form a proportion.**

13) $\frac{5}{12}$ and $\frac{15}{36}$

14) $\frac{2}{4}$ and $\frac{18}{36}$

15) $\frac{7}{8}$ and $\frac{28}{32}$

16) $\frac{3}{8}$ and $\frac{27}{64}$

17) $\frac{1}{14}$ and $\frac{5}{65}$

18) $\frac{7}{11}$ and $\frac{70}{100}$

19) $\frac{12}{15}$ and $\frac{48}{60}$

20) $\frac{3}{17}$ and $\frac{36}{204}$

21) $\frac{1.2}{1.5}$ and $\frac{1.44}{22.5}$

22) $\frac{1.3}{1.1}$ and $\frac{3.9}{33}$

23) $\frac{0.7}{0.9}$ and $\frac{6.3}{8.1}$

24) $\frac{2.4}{3.2}$ and $\frac{48}{64}$

🍃 **Calculate each proportion.**

25) $\frac{14}{16} = \frac{21}{x}, x = \_\_\_$

26) $\frac{3}{28} = \frac{42}{x}, x = \_\_\_$

27) $\frac{19}{5} = \frac{38}{x}, x = \_\_\_$

28) $\frac{3}{10} = \frac{x}{140}, x = \_\_\_$

29) $\frac{4}{9} = \frac{x}{108}, x = \_\_\_$

30) $\frac{7}{32} = \frac{21}{x}, x = \_\_\_$

31) $\frac{9}{8} = \frac{108}{x}, x = \_\_\_$

32) $\frac{12}{17} = \frac{48}{x}, x = \_\_\_$

33) $\frac{1.4}{5} = \frac{x}{30}, x = \_\_\_$

34) $\frac{1.6}{12} = \frac{x}{60}, x = \_\_\_$

35) $\frac{3.5}{15} = \frac{x}{315}, x = \_\_\_$

36) $\frac{4.7}{2.5} = \frac{x}{50}, x = \_\_\_$

Similarity and Ratios

✍ **Each pair of figures is similar. Find the missing side.**

1)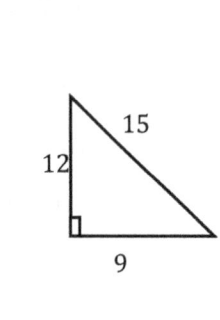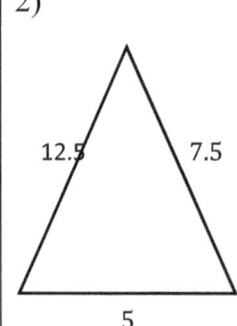

2)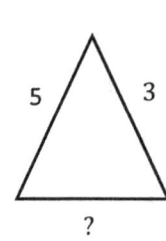

3)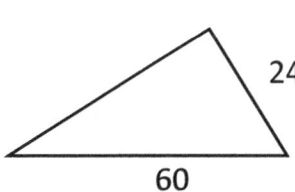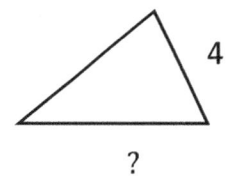

4)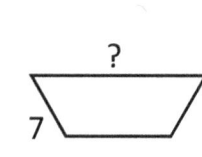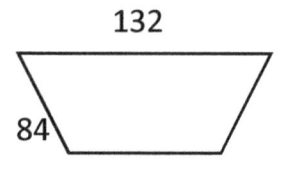

✍ **Calculate.**

5) Two rectangles are similar. The first is 14 feet wide and 70 feet long. The second is 30 feet wide. What is the length of the second rectangle?
_____

6) Two rectangles are similar. One is 3.2 meters by 15 meters. The longer side of the second rectangle is 42 meters. What is the other side of the second rectangle? _____

7) A building casts a shadow 24 ft long. At the same time a girl 10 ft tall casts a shadow 6 ft long. How tall is the building? _____

8) The scale of a map of Texas is 8 inches: 52 miles. If you measure the distance from Dallas to Martin County as 28.8 inches, approximately how far is Martin County from Dallas? _____

Ratio and Rates Word Problems

✎ **Find the answer for each word problem.**

1) Mason has 32 red cards and 40 green cards. What is the ratio of Mason 's red cards to his green cards? _____

2) In a party, 24 soft drinks are required for every 42 guests. If there are 378 guests, how many soft drinks is required? _____

3) In Mason's class, 54 of the students are tall and 30 are short. In Michael's class 126 students are tall and 70 students are short. Which class has a higher ratio of tall to short students? _____

4) The price of 4 apples at the Quick Market is $3.65. The price of 6 of the same apples at Walmart is $4.25. Which place is the better buy? _____

5) The bakers at a Bakery can make 90 bagels in 3 hours. How many bagels can they bake in 17 hours? What is that rate per hour? _____

6) You can buy 8 cans of green beans at a supermarket for $5.60. How much does it cost to buy 56 cans of green beans? _____

7) The ratio of boys to girls in a class is 4: 7. If there are 16 boys in the class, how many girls are in that class? _____

8) The ratio of red marbles to blue marbles in a bag is 3: 4. If there are 42 marbles in the bag, how many of the marbles are red? _____

Percentage Calculations

✎ **Calculate the given percent of each value.**

1) 3% of 60 = ____
2) 20% of 80 = ____
3) 25% of 80 = ____
4) 24% of 50 = ____
5) 18% of 150 = ____
6) 70% of 35 = ____

7) 15% of 28 = ____
8) 32% of 300 = ____
9) 54% of 80 = ____
10) 10% of 610 = ____
11) 35% of 520 = ____
12) 64% of 110 = ____

13) 44% of 200 = ____
14) 28% of 94 = ____
15) 30% of 85 = ____
16) 68% of 102 = ____
17) 45% of 160 = ____
18) 55% of 220 = ____

✎ **Calculate the percent of each given value.**

19) ____% of 18 = 9
20) ____% of 50 = 40
21) ____% of 140 = 7
22) ____% of 158 = 39.5
23) ____% of 75 = 9.375

24) ____% of 45 = 11.25
25) ____% of 90 = 22.5
26) ____% of 650 = 19.5
27) ____% of 480 = 24
28) ____% of 400 = 57.32

✎ **Calculate each percent problem.**

29) A Cinema has 132 seats. 92 seats were sold for the current movie. What percent of seats are empty? ____ %

30) There are 52 boys and 68 girls in a class. 55.00% of the students in the class take the bus to school. How many students do not take the bus to school? ____

Percent Problems

✎ **Calculate each problem.**

1) 30 is what percent of 60? ___%

2) 32 is what percent of 80? ___%

3) 72 is what percent of 45? ___%

4) 8 is what percent of 200? ___%

5) 9 is what percent of 600? ___%

6) 30 is what percent of 500? ___%

7) 70 is what percent of 350? ___%

8) 44 is what percent of 550? ___%

9) 270 is what percent of 900? ___%

10) 180 is what percent of 720? ___%

11) 37.5 is what percent of 75? ___%

12) 27.5 is what percent of 55? ___%

13) 60 is what percent of 750? ___%

14) 22.5 is what percent of 18? ___%

15) 36 is what percent of 24? ___%

16) 18 is what percent of 60? ___%

17) 140 is what percent of 280? ___%

18) 128 is what percent of 40? ___%

✎ **Calculate each percent word problem.**

19) There are 32 employees in a company. On a certain day, 24 were present. What percent showed up for work? ____%

20) A metal bar weighs 36 ounces. 40% of the bar is gold. How many ounces of gold are in the bar? _____

21) A crew is made up of 12 women; the rest are men. If 20% of the crew are women, how many people are in the crew? _____

22) There are 32 students in a class and 8 of them are girls. What percent are boys? ____%

23) The Royals softball team played 310 games and won 248 of them. What percent of the games did they lose? ____%

Discount, Tax and Tip

✎ Find the selling price of each item.

1) Original price of a computer: $450
   Tax: 8%   Selling price: $_____

2) Original price of a laptop: $240
   Tax: 4%   Selling price: $_____

3) Original price of a sofa: $900
   Tax: 12%   Selling price: $_____

4) Original price of a car: $10,400
   Tax: 2.5%   Selling price: $_____

5) Original price of a Table: $400
   Tax: 3%   Selling price: $_____

6) Original price of a house: $360,000
   Tax: 2.8%   Selling price: $_____

7) Original price of a tablet: $150
   Discount: 24%   Selling price: $____

8) Original price of a chair: $180
   Discount: 20%   Selling price: $____

9) Original price of a book: $80
   Discount: 30%   Selling price: $____

10) Original price of a cellphone: $800
    Discount: 20%   Selling price: $___

11) Food bill: $56
    Tip: 15%   Price: $_____

12) Food bill: $50
    Tipp: 10%   Price: $_____

13) Food bill: $94
    Tip: 25%   Price: $_____

14) Food bill: $48
    Tipp: 30%   Price: $_____

✎ Find the answer for each word problem.

15) Nicolas hired a moving company. The company charged $400 for its services, and Nicolas gives the movers a 20% tip. How much does Nicolas tip the movers? $_____

16) Mason has lunch at a restaurant and the cost of his meal is $80. Mason wants to leave a 20% tip. What is Mason's total bill including tip? $_____

17) The sales tax in Texas is 14.45% and an item costs $300. How much is the tax? $_____

18) The price of a table at Best Buy is $520. If the sales tax is 4%, what is the final price of the table including tax? $_____

Percent of Change

✎ Find each percent of change.

1) From 300 to 600. ___ %

2) From 45 ft to 225 ft. ___ %

3) From $60 to $420. ___ %

4) From 30 cm to 120 cm. ___ %

5) From 10 to 30. ___ %

6) From 12 to 30. ___ %

7) From 140 to 210. ___ %

8) From 800 to 400. ___ %

9) From 85 to 51. ___ %

10) From 152 to 76. ___ %

✎ Calculate each percent of change word problem.

11) Bob got a raise, and his hourly wage increased from $32 to $40. What is the percent increase? ___ %

12) The price of a pair of shoes increases from $70 to $112. What is the percent increase? ___ %

13) At a coffee shop, the price of a cup of coffee increased from $1.90 to $2.28. What is the percent increase in the cost of the coffee? ___ %

14) 30 cm are cut from a 120 cm board. What is the percent decrease in length? ___ %

15) In a class, the number of students has been increased from 54 to 81. What is the percent increase? ___ %

16) The price of gasoline rose from $22.4 to $25.76 in one month. By what percent did the gas price rise? ___ %

17) A shirt was originally priced at $19. It went on sale for $22.80. What was the percent that the shirt was discounted? ___ %

Simple Interest

✎ Determine the simple interest for these loans.

1) $210 at 15% for 4 years. $ _____

2) $1,200 at 6% for 3 years. $ _____

3) $950 at 25% for 2 years. $ _____

4) $6,500 at 1.5% for 7 months. $ _____

5) $240 at 5% for 8 months. $ _____

6) $28,000 at 3.5% for 6 years. $ _____

7) $9,600 at 8% for 2 years. $ _____

8) $500 at 4.2% for 5 years. $ _____

9) $700 at 2.8 % for 6 months. $ _____

10) $9,000 at 1.6% for 4 years. $ _____

✎ Calculate each simple interest word problem.

11) A new car, valued at $16,000, depreciates at 3.5% per year. What is the value of the car two year after purchase? $_____

12) Sara puts $9,000 into an investment yielding 8% annual simple interest; she left the money in for three years. How much interest does Sara get at the end of those three years? $_____

13) A bank is offering 12.5% simple interest on a savings account. If you deposit $32,400, how much interest will you earn in one years? $_____

14) $2,400 interest is earned on a principal of $10,000 at a simple interest rate of 12% interest per year. For how many years was the principal invested? _____

15) In how many years will $1,200 yield an interest of $384 at 8% simple interest? _____

16) Jim invested $5,000 in a bond at a yearly rate of 2.5%. He earned $375 in interest. How long was the money invested? _____

Answers of Worksheets – Chapter 3

Simplifying Ratios

1) 3:4
2) 1:10
3) 4:7
4) 1:3
5) 1:10
6) 1:8
7) 2:8
8) 2:5
9) 1:6
10) 7:9
11) 2:3
12) 7:2
13) 10:1
14) 3:4
15) 1:8
16) 5:7
17) 7:9
18) 3:1
19) 1:3
20) 7:1
21) 5:1
22) 5:4
23) 1:2
24) 1:30
25) $\frac{1}{2}$
26) $\frac{3}{5}$
27) $\frac{3}{7}$
28) $\frac{1}{3}$
29) $\frac{1}{3}$
30) $\frac{3}{14}$
31) $\frac{1}{2}$
32) $\frac{1}{5}$
33) $\frac{5}{12}$
34) $\frac{2}{9}$
35) $\frac{11}{13}$
36) $\frac{4}{9}$
37) $\frac{1}{12}$
38) $\frac{3}{11}$
39) $\frac{1}{15}$
40) $\frac{1}{4}$
41) $\frac{4}{3}$
42) $\frac{1}{5}$
43) $\frac{22}{41}$
44) $\frac{1}{9}$
45) $\frac{3}{5}$

Proportional Ratios

1) 12
2) 90
3) 5
4) 24
5) 63
6) 14
7) 23
8) 8
9) 48
10) 50
11) 14.2
12) 18
13) Yes
14) Yes
15) Yes
16) No
17) No
18) No
19) Yes
20) Yes
21) No
22) No
23) Yes
24) Yes
25) 24
26) 392
27) 10
28) 42
29) 48
30) 96
31) 96
32) 68
33) 8.4
34) 8
35) 73.5
36) 94

Similarity and ratios

1) 30
2) 2
3) 10
4) 11
5) 150 feet
6) 8.96 meters
7) 40 feet
8) 187.2 miles

Ratio and Rates Word Problems

1) 4:5
2) 252

3) The ratio for both classes is 9 to 5.
4) Walmart is a better buy.
5) 510, the rate is 30 per hour.
6) $39.20
7) 28
8) 18

Percentage Calculations

1) 1.8
2) 1.6
3) 20
4) 12
5) 27
6) 24.5
7) 4.2
8) 96
9) 43.2
10) 61
11) 182
12) 70.4
13) 88
14) 26.32
15) 25.5
16) 69.36
17) 72
18) 121
19) 50%
20) 80%
21) 5%
22) 25%
23) 12.5%
24) 25%
25) 25%
26) 3%
27) 5%
28) 14.33%
29) 30.30%
30) 54

Percent Problems

1) 50%
2) 40%
3) 160%
4) 4%
5) 1.5%
6) 6%
7) 20%
8) 8%
9) 30%
10) 25%
11) 50%
12) 50%
13) 8%
14) 125%
15) 150%
16) 30%
17) 50%
18) 320%
19) 75%
20) 14.4 ounces
21) 60
22) 75%
23) 20%

Discount, Tax and Tip

1) $486.00
2) $249.60
3) $1,008.00
4) $10,660.00
5) $412.00
6) $370,080
7) $144.00
8) $144.00
9) $56.00
10) $640.00
11) $64.40
12) $55.00
13) $117.50
14) $62.40
15) $80.00
16) $96.00
17) $43.35
18) $540.80

Percent of Change

1) 100%	7) 50%	13) 20%
2) 400%	8) 50%	14) 25%
3) 600%	9) 40%	15) 50%
4) 300%	10) 50%	16) 15%
5) 200%	11) 25%	17) 20%
6) 150%	12) 60%	

Simple Interest

1) $126.00	7) $1,536.00	13) $4,050.00
2) $216.00	8) $105.00	14) 2 years
3) $475.00	9) $9.80	15) 4 years
4) $56.875	10) $576.00	16) 3 years
5) $8.00	11) $14,880.00	
6) $5,880.00	12) $2,160.00	

Chapter 4: Exponents

Topics that you will practice in this chapter:

- ✓ Multiplication Property of Exponents
- ✓ Zero and Negative Exponents
- ✓ Division Property of Exponents
- ✓ Powers of Products and Quotients
- ✓ Negative Exponents and Negative Bases
- ✓ Scientific Notation
- ✓ Square Roots
- ✓ Simplifying Radical Expressions

Mathematics is no more computation than typing is literature.
– John Allen Paulos

Multiplication Property of Exponents

✏ Simplify and write the answer in exponential form.

1) $2 \times 2^5 =$

2) $7^2 \times 7 =$

3) $8^3 \times 8^3 =$

4) $9^4 \times 9^3 =$

5) $4^2 \times 4^4 \times 4 =$

6) $5 \times 5^2 \times 5^3 =$

7) $9^3 \times 9^3 \times 9 \times 9 =$

8) $4x \times x =$

9) $x^5 \times x^3 =$

10) $x^6 \times x^2 =$

11) $x^2 \times x^4 \times x^5 =$

12) $7x \times 7x =$

13) $4x^2 \times 5x^3 =$

14) $12x^3 \times x =$

15) $3x^2 \times 3x^2 \times 3x^2 =$

16) $7x^5 \times 2x^3 =$

17) $x^8 \times 2x =$

18) $3x \times 3x^3 =$

19) $6x^2 \times 2x^5 =$

20) $3yx^3 \times 12x =$

21) $8x^3 \times y^5 x^2 =$

22) $4y^7 x^2 \times 3y^2 x^5 =$

23) $9yx^2 \times 4x^5 y^2 =$

24) $10x^4 \times 11x^4 y^4 =$

25) $9x^3 y^4 \times 9x^6 y^2 =$

26) $12x^4 y^4 \times 6xy^3 =$

27) $9xy^4 \times 11x^3 y^3 =$

28) $6x^2 y^4 \times 8x^3 y^6 =$

29) $8x \times y^7 x^2 \times 5y^3 =$

30) $3yx^3 \times 2y^3 x^2 \times 7xy =$

31) $8yx^5 \times 3y^4 x \times 3xy^3 =$

32) $9x^3 \times 11y^4 x^3 \times 2yx^4 =$

Zero and Negative Exponents

✏️ **Evaluate the following expressions.**

1) $1^{-5} =$

2) $2^{-4} =$

3) $2^{-5} =$

4) $3^0 =$

5) $3^{-2} =$

6) $2^{-7} =$

7) $13^{-2} =$

8) $14^{-2} =$

9) $2^{-8} =$

10) $20^{-2} =$

11) $19^{-1} =$

12) $3^{-6} =$

13) $15^{-2} =$

14) $10^{-2} =$

15) $16^{-2} =$

16) $30^{-2} =$

17) $8^{-4} =$

18) $3^{-7} =$

19) $2^{-10} =$

20) $10^{-3} =$

21) $18^{-2} =$

22) $25^{-2} =$

23) $40^{-2} =$

24) $50^{-2} =$

25) $11^{-3} =$

26) $22^{-2} =$

27) $17^{-2} =$

28) $3^{-8} =$

29) $4^{-5} =$

30) $60^{-2} =$

31) $(\frac{1}{3})^{-2}$

32) $(\frac{1}{5})^{-3} =$

33) $(\frac{1}{8})^{-2} =$

34) $(\frac{2}{5})^{-2} =$

35) $(\frac{1}{15})^{-2} =$

36) $(\frac{7}{12})^{-2} =$

37) $(\frac{1}{20})^{-2} =$

38) $(\frac{1}{7})^{-3} =$

39) $(\frac{2}{3})^{-5} =$

40) $(\frac{9}{11})^{-1} =$

41) $(\frac{8}{9})^{-2} =$

42) $(\frac{1}{8})^{-3} =$

Division Property of Exponents

✎ Simplify.

1) $\dfrac{5^2}{5^6} =$

2) $\dfrac{6^9}{6^5} =$

3) $\dfrac{9^7}{9} =$

4) $\dfrac{3}{3^3} =$

5) $\dfrac{2x}{x^8} =$

6) $\dfrac{4 \times 4^5}{4^5 \times 4^2} =$

7) $\dfrac{12^6}{12^2} =$

8) $\dfrac{7 \times 7^9}{7^2 \times 7^4} =$

9) $\dfrac{4^5 \times 4^8}{4^2 \times 4^{11}} =$

10) $\dfrac{20x}{40x^4} =$

11) $\dfrac{8x^9}{9x^6} =$

12) $\dfrac{24x^3}{16x^5} =$

13) $\dfrac{25x^2}{50y^8} =$

14) $\dfrac{60xy^5}{12x^4y^2} =$

15) $\dfrac{8x^7}{12x} =$

16) $\dfrac{48x^2y^4}{16x^5} =$

17) $\dfrac{50x^6}{25x^9y^{14}} =$

18) $\dfrac{90yx^8}{15yx^9} =$

19) $\dfrac{18x^9y}{36x^{12}y^3} =$

20) $\dfrac{9x^8}{81x^8} =$

21) $\dfrac{9x^{-7}}{11x^{-3}} =$

Powers of Products and Quotients

✏️ **Simplify.**

1) $(4^2)^3 =$

2) $(5^2)^2 =$

3) $(3 \times 3^2)^3 =$

4) $(3 \times 2^3)^2 =$

5) $(15^2 \times 15^2)^5 =$

6) $(7^2 \times 7^3)^4 =$

7) $(9 \times 9^2)^2 =$

8) $(4^6)^3 =$

9) $(7x^7)^3 =$

10) $(8x^4y^3)^2 =$

11) $(3x^3y^2)^4 =$

12) $(4x^2y^2)^2 =$

13) $(3x^5y^2)^3 =$

14) $(4x^3y^2)^3 =$

15) $(2x^3x)^5 =$

16) $(6x^4x^2)^2 =$

17) $(7x^{12}y^5)^2 =$

18) $(5x^7x^4)^3 =$

19) $(8x^2 \times 6x)^2 =$

20) $(9x^{14}y^3)^3 =$

21) $(5x^4y^2)^4 =$

22) $(3x^3y^7)^5 =$

23) $(8x \times 2y^3)^2 =$

24) $\left(\frac{8x}{x^3}\right)^3 =$

25) $\left(\frac{x^4y^5}{x^3y^5}\right)^7 =$

26) $\left(\frac{36xy}{6x^5}\right)^2 =$

27) $\left(\frac{x^4}{x^5y^2}\right)^3 =$

28) $\left(\frac{xy^2}{x^3y^8}\right)^{-3} =$

29) $\left(\frac{5xy^7}{x^2}\right)^3 =$

30) $\left(\frac{xy^5}{2xy^3}\right)^{-6} =$

Negative Exponents and Negative Bases

✏️ **Simplify.**

1) $-4^{-2} =$

2) $-7^{-1} =$

3) $-5^{-2} =$

4) $-x^{-9} =$

5) $10x^{-2} =$

6) $-7x^{-4} =$

7) $-15x^{-4} =$

8) $-15x^{-7}y^{-4} =$

9) $32x^{-9}y^{-3} =$

10) $45a^{-7}b^{-3} =$

11) $-25x^3y^{-5} =$

12) $-\dfrac{18}{x^{-9}} =$

13) $-\dfrac{13x}{a^{-8}} =$

14) $\left(-\dfrac{1}{3}\right)^{-4} =$

15) $\left(-\dfrac{3}{4}\right)^{-3} =$

16) $-\dfrac{12}{a^{-6}b^{-4}} =$

17) $-\dfrac{48x}{x^{-6}} =$

18) $-\dfrac{a^{-12}}{b^{-5}} =$

19) $-\dfrac{27}{x^{-5}} =$

20) $\dfrac{12b}{-48c^{-6}} =$

21) $\dfrac{24ab}{a^{-4}b^{-3}} =$

22) $-\dfrac{8n^{-7}}{40p^{-9}} =$

23) $\dfrac{9ab^{-6}}{-5c^{-2}} =$

24) $\left(\dfrac{2a}{3c}\right)^{-4} =$

25) $\left(-\dfrac{8x}{5yz}\right)^{-2} =$

26) $\dfrac{9ab^{-6}}{-4c^{-3}} =$

27) $\left(-\dfrac{x^3}{x^4}\right)^{-5} =$

28) $\left(-\dfrac{x^{-3}}{3x^3}\right)^{-3} =$

29) $\left(-\dfrac{x^{-6}}{x^4}\right)^{-3} =$

Scientific Notation

✎ Write each number in scientific notation.

1) $0.226 =$

2) $0.05 =$

3) $4.8 =$

4) $90 =$

5) $120 =$

6) $0.123 =$

7) $82 =$

8) $5,400 =$

9) $2,460 =$

10) $75,300 =$

11) $61,000,000 =$

12) $0.00009 =$

13) $468,000 =$

14) $0.00458 =$

15) $0.000087 =$

16) $31,800,000 =$

17) $950,000 =$

18) $9,000,000,000 =$

19) $0.0007 =$

20) $0.00041 =$

✎ Write each number in standard notation.

21) $4 \times 10^{-2} =$

22) $7 \times 10^{-4} =$

23) $4.3 \times 10^6 =$

24) $7 \times 10^{-4} =$

25) $8.7 \times 10^{-3} =$

26) $12 \times 10^5 =$

27) $35 \times 10^3 =$

28) $1.89 \times 10^5 =$

29) $13 \times 10^{-6} =$

30) $7.3 \times 10^{-4} =$

Answers of Worksheets – Chapter 4

Multiplication Property of Exponents

1) 2^6
2) 7^3
3) 8^6
4) 9^7
5) 4^7
6) 5^6
7) 9^8
8) $4x^2$
9) x^8
10) x^8
11) x^{11}
12) $49x^2$
13) $20x^5$
14) $12x^4$
15) $27x^6$
16) $14x^8$
17) $2x^8$
18) $9x^4$
19) $12x^7$
20) $36x^4y$
21) $8x^5y^5$
22) $12x^7y^9$
23) $36x^7y^3$
24) $110x^8y^4$
25) $81x^9y^6$
26) $72x^5y^7$
27) $99x^4y^7$
28) $48x^5y^{10}$
29) $40x^3y^{10}$
30) $42x^6y^5$
31) $72x^7y^8$
32) $198x^{10}y^5$

Zero and Negative Exponents

1) 1
2) $\frac{1}{16}$
3) $\frac{1}{32}$
4) 1
5) $\frac{1}{9}$
6) $\frac{1}{128}$
7) $\frac{1}{169}$
8) $\frac{1}{196}$
9) $\frac{1}{256}$
10) $\frac{1}{400}$
11) $\frac{1}{19}$
12) $\frac{1}{729}$
13) $\frac{1}{225}$
14) $\frac{1}{100}$
15) $\frac{1}{256}$
16) $\frac{1}{900}$
17) $\frac{1}{4,096}$
18) $\frac{1}{2,187}$
19) $\frac{1}{1,024}$
20) $\frac{1}{1,000}$
21) $\frac{1}{324}$
22) $\frac{1}{625}$
23) $\frac{1}{1,600}$
24) $\frac{1}{2,500}$
25) $\frac{1}{1,331}$
26) $\frac{1}{484}$
27) $\frac{1}{289}$
28) $\frac{1}{6,561}$
29) $\frac{1}{1,024}$
30) $\frac{1}{3,600}$
31)
32)
33)
34) 6.25
35) 225
36) $\frac{144}{49}$
37) 400
38) 343
39) $\frac{243}{32}$
40) $\frac{11}{9}$
41) $\frac{81}{64}$
42) 512

Division Property of Exponents

1) $\frac{1}{5^4}$
2) 6^4
3) 9^6
4) $\frac{1}{3^2}$
5) $\frac{2}{x^7}$
6) $\frac{1}{4}$
7) 12^4
8) 7^4
9) 1
10) $\frac{1}{2x^3}$
11) $\frac{8x^3}{9}$
12) $\frac{3}{2x^2}$
13) $\frac{x^2}{2y^8}$
14) $\frac{5y^3}{x^3}$
15) $\frac{2x^6}{3}$
16) $\frac{3y^4}{x^3}$
17) $\frac{2}{x^3y^{14}}$

CLEP College Math Workbook

18) $\frac{6}{x}$ 19) $\frac{1}{2x^3y^2}$ 20) $\frac{1}{9}$ 21) $\frac{9}{11x^4}$

Powers of Products and Quotients

1) 4^6
2) 5^4
3) 3^9
4) 24^2
5) 15^{20}
6) 7^{20}
7) 9^6
8) 4^{18}
9) $343x^{21}$
10) $64x^8y^6$
11) $81x^{12}y^8$
12) $16x^4y^4$
13) $27x^{15}y^6$
14) $64x^9y^6$
15) $32x^{20}$
16) $36x^{12}$
17) $49x^{24}y^{10}$
18) $125x^{33}$
19) $2,304x^6$
20) $729x^{42}y^9$
21) $625x^{16}y^9$
22) $243x^{15}y^8$
23) $256x^2y^6$
24) $\frac{512}{x^6}$
25) x^7
26) $\frac{36y^2}{x^8}$
27) $\frac{1}{x^3y^6}$
28) x^6y^{18}
29) $\frac{125y^{21}}{x^3}$
30) $\frac{64}{y^{12}}$

Negative Exponents and Negative Bases

1) $-\frac{1}{16}$
2) $-\frac{1}{7}$
3) $-\frac{1}{25}$
4) $-\frac{1}{x^9}$
5) $\frac{10}{x^2}$
6) $-\frac{7}{x^4}$
7) $-\frac{15}{x^4}$
8) $-\frac{15}{x^7y^4}$
9) $\frac{32}{x^9y^3}$
10) $\frac{45}{a^7b^3}$
11) $-\frac{25x^3}{y^5}$
12) $-18x^9$
13) $-13xa^8$
14) 81
15) $-\frac{64}{27}$
16) $-12a^6b^4$
17) $-48x^7$
18) $-\frac{b^5}{a^{12}}$
19) $-27x^5$
20) $-\frac{bc^6}{4}$
21) $24a^5b^4$
22) $-\frac{p^9}{5n^7}$
23) $-\frac{9ac^2}{5b^6}$
24) $\frac{81c^4}{16a^4}$
25) $\frac{25y^2z^2}{64x^2}$
26) $-\frac{9ac^3}{4b^6}$
27) $-x^5$
28) $-27x^{18}$
29) $-x^{30}$

Scientific Notation

1) 2.26×10^{-1}
2) 5×10^{-2}
3) 4.8×10^0
4) 9×10^1
5) 1.2×10^2
6) 1.23×10^{-1}
7) 8.2×10^1
8) 5.4×10^3
9) 2.46×10^3

10) 7.53×10^4
11) 61×10^6
12) 9×10^{-5}
13) 4.68×10^5
14) 4.58×10^{-3}
15) 8.7×10^{-5}
16) 3.18×10^7

17) 9.5×10^5
18) 9×10^9
19) 7×10^{-4}
20) 4.1×10^{-4}
21) 0.04
22) 0.0007
23) 4,300,000

24) 0.0007
25) 0.0087
26) 1,200,000
27) 35,000
28) 189,000
29) 0.000013
30) 0.00073

Chapter 5:
Radicals Expressions

Topics that you will practice in this chapter:

- ✓ Square Roots
- ✓ Simplifying Radical Expressions
- ✓ Simplifying Radical Expressions Involving Fractions
- ✓ Multiplying Radical Expressions
- ✓ Adding and Subtracting Radical Expressions

Mathematics is no more computation than typing is literature.
  – John Allen Paulos

Square Roots

✏️ **Find the value each square root.**

1) $\sqrt{64} = $ ____

2) $\sqrt{4} = $ ____

3) $\sqrt{289} = $ ____

4) $\sqrt{0.25} = $ ____

5) $\sqrt{0.01} = $ ____

6) $\sqrt{0.09} = $ ____

7) $\sqrt{1,600} = $ ____

8) $\sqrt{2.25} = $ ____

9) $\sqrt{0} = $ ____

10) $\sqrt{0.04} = $ ____

11) $\sqrt{0.36} = $ ____

12) $\sqrt{0.81} = $ ____

13) $\sqrt{0.49} = $ ____

14) $\sqrt{1.21} = $ ____

15) $\sqrt{1.69} = $ ____

16) $\sqrt{0.16} = $ ____

17) $\sqrt{529} = $ ____

18) $\sqrt{625} = $ ____

19) $\sqrt{0.81} = $ ____

20) $\sqrt{20} = $ ____

21) $\sqrt{50} = $ ____

22) $\sqrt{676} = $ ____

23) $\sqrt{270} = $ ____

24) $\sqrt{32} = $ ____

✏️ **Evaluate.**

25) $\sqrt{4} \times \sqrt{16} = $ _____

26) $\sqrt{49} \times \sqrt{64} = $ _____

27) $\sqrt{2} \times \sqrt{8} = $ _____

28) $\sqrt{17} \times \sqrt{17} = $ _____

29) $\sqrt{13} \times \sqrt{13} = $ _____

30) $\sqrt{15} \times \sqrt{15} = $ _____

31) $\sqrt{19} + \sqrt{19} = $ _____

32) $\sqrt{1} + \sqrt{1} = $ _____

33) $8\sqrt{7} - 2\sqrt{7} = $ _____

34) $7\sqrt{10} \times 6\sqrt{10} = $ _____

35) $9\sqrt{5} \times 2\sqrt{5} = $ _____

36) $8\sqrt{3} - \sqrt{12} = $ _____

Simplifying Radical Expressions

✎ **Simplify.**

1) $\sqrt{13y^2} =$

2) $\sqrt{60x^3} =$

3) $\sqrt[3]{27a} =$

4) $\sqrt{81x^2} =$

5) $\sqrt{150a} =$

6) $\sqrt[3]{135w^3} =$

7) $\sqrt{200x} =$

8) $\sqrt{192v} =$

9) $\sqrt[3]{64x} =$

10) $\sqrt{84x^3} =$

11) $\sqrt{121x^2} =$

12) $\sqrt[3]{48a} =$

13) $\sqrt{480} =$

14) $\sqrt{1,575p^2} =$

15) $\sqrt{108m^6} =$

16) $\sqrt{198x^3y^2} =$

17) $\sqrt{169x^2y^3} =$

18) $\sqrt{25a^6} =$

19) $\sqrt{50x^2y^3} =$

20) $\sqrt[3]{512y^3} =$

21) $2\sqrt{144x^2} =$

22) $3\sqrt{400x^2} =$

23) $\sqrt[3]{189xy^4} =$

24) $\sqrt[3]{1,331x^3y^5} =$

25) $3\sqrt{150a} =$

26) $\sqrt[3]{729y} =$

27) $3\sqrt{18xyr^3} =$

28) $6\sqrt{225x^2yz^6} =$

29) $3\sqrt[3]{125x^3y^2} =$

30) $7\sqrt{12a^2bc^4} =$

31) $4\sqrt[3]{1,000x^9y^{15}} =$

Multiplying Radical Expressions

✍ **Simplify.**

1) $\sqrt{11} \times \sqrt{11} =$

2) $\sqrt{5} \times \sqrt{15} =$

3) $\sqrt{3} \times \sqrt{12} =$

4) $\sqrt{20} \times \sqrt{25} =$

5) $\sqrt{5} \times (-2)\sqrt{35} =$

6) $2\sqrt{12} \times \sqrt{3} =$

7) $4\sqrt{24} \times \sqrt{6} =$

8) $\sqrt{5} \times (-\sqrt{75}) =$

9) $\sqrt{88} \times \sqrt{40} =$

10) $2\sqrt{45} \times 4\sqrt{105} =$

11) $\sqrt{32}(2 + \sqrt{2}) =$

12) $\sqrt{13x^2} \times \sqrt{13x} =$

13) $-7\sqrt{12} \times \sqrt{3} =$

14) $5\sqrt{19x^3} \times \sqrt{19x^3} =$

15) $\sqrt{15x^2} \times \sqrt{5x} =$

16) $-8\sqrt{2x} \times \sqrt{6x^5} =$

17) $-4\sqrt{5x} \times 5\sqrt{45x^2} =$

18) $-3\sqrt{27}(3 + \sqrt{15}) =$

19) $\sqrt{8x}(3 - \sqrt{2x}) =$

20) $\sqrt{5x}(10\sqrt{5x} + \sqrt{40}) =$

21) $\sqrt{18r}(6 + \sqrt{6}) =$

22) $-12\sqrt{3x} \times 3\sqrt{15x^3} =$

23) $-4\sqrt{27x} \times 6\sqrt{3x} =$

24) $-3\sqrt{10v^2}(-3\sqrt{15v}) =$

25) $(\sqrt{8} - 3)(\sqrt{8} + \sqrt{9}) =$

26) $(-3\sqrt{5} + 7)(\sqrt{5} \quad 3) =$

27) $(3 - 4\sqrt{5})(-2 + \sqrt{4}) =$

28) $(13 - 2\sqrt{5})(3 - \sqrt{5}) =$

29) $(5 - \sqrt{3x})(5 + \sqrt{3x}) =$

30) $(-6 + 3\sqrt{3r})(-6 + \sqrt{3r}) =$

31) $(-\sqrt{5n} + 8)(-\sqrt{5} - 8) =$

32) $(-3 + 3\sqrt{2})(3 - 2\sqrt{2x}) =$

Simplifying Radical Expressions Involving Fractions

✎ **Simplify.**

1) $\dfrac{\sqrt{3}}{\sqrt{2}} =$

2) $\dfrac{\sqrt{24}}{\sqrt{40}} =$

3) $\dfrac{\sqrt{12}}{2\sqrt{6}} =$

4) $\dfrac{21}{\sqrt{5}} =$

5) $\dfrac{15\sqrt{8r}}{\sqrt{m^5}} =$

6) $\dfrac{8\sqrt{2}}{\sqrt{m}} =$

7) $\dfrac{15\sqrt{25n^2}}{5\sqrt{15n}} =$

8) $\dfrac{\sqrt{8x^3y^5}}{\sqrt{2y^2x^4}} =$

9) $\dfrac{2}{2+\sqrt{5}} =$

10) $\dfrac{2-12\sqrt{x}}{\sqrt{24x}} =$

11) $\dfrac{2\sqrt{x}}{\sqrt{x}-\sqrt{y}} =$

12) $\dfrac{3-\sqrt{5}}{5-\sqrt{3}} =$

13) $\dfrac{5+\sqrt{12}}{5-\sqrt{3}} =$

14) $\dfrac{8}{-3-3\sqrt{3}} =$

15) $\dfrac{5}{2+\sqrt{15}} =$

16) $\dfrac{\sqrt{11}-\sqrt{7}}{\sqrt{7}-\sqrt{11}} =$

17) $\dfrac{\sqrt{8}+\sqrt{2}}{\sqrt{2}-\sqrt{8}} =$

18) $\dfrac{2\sqrt{2}-\sqrt{3}}{3\sqrt{2}+\sqrt{3}} =$

19) $\dfrac{\sqrt{11}+3\sqrt{5}}{3-\sqrt{11}} =$

20) $\dfrac{\sqrt{7}+\sqrt{13}}{13-\sqrt{7}} =$

21) $\dfrac{\sqrt{125a^7b^5}}{\sqrt{5ab^4}} =$

22) $\dfrac{72\sqrt{24m^3}}{9\sqrt{m}} =$

Adding and Subtracting Radical Expressions

✎ **Simplify.**

1) $\sqrt{5} + \sqrt{20} =$

2) $7\sqrt{44} + 7\sqrt{11} =$

3) $9\sqrt{3} - 3\sqrt{12} =$

4) $8\sqrt{9} - 5\sqrt{3} =$

5) $9\sqrt{80} - 9\sqrt{20} =$

6) $-\sqrt{32} - 5\sqrt{8} =$

7) $-12\sqrt{16} - 9\sqrt{64} =$

8) $15\sqrt{8} + 6\sqrt{32} =$

9) $16\sqrt{9} - 12\sqrt{36} =$

10) $-7\sqrt{7} + 11\sqrt{63} =$

11) $-24\sqrt{13} + 18\sqrt{117} =$

12) $25\sqrt{5} - 12\sqrt{45} =$

13) $-8\sqrt{99} + 2\sqrt{11} =$

14) $6\sqrt{3} - 2\sqrt{12} =$

15) $8\sqrt{20} + 3\sqrt{5} =$

16) $5\sqrt{28} - 8\sqrt{63} =$

17) $\sqrt{144} - \sqrt{121} =$

18) $6\sqrt{18} - 2\sqrt{2} =$

19) $-12\sqrt{7} + 21\sqrt{28} =$

20) $5\sqrt{60} - 5\sqrt{15} =$

21) $6\sqrt{54} - 3\sqrt{6} =$

22) $-4\sqrt{3} + 8\sqrt{75} =$

23) $-9\sqrt{20} - 7\sqrt{5} =$

24) $-\sqrt{216x} + 6\sqrt{6x} =$

25) $\sqrt{14y^2} + y\sqrt{126} =$

26) $\sqrt{11mn^2} + n\sqrt{99m} =$

27) $-8\sqrt{48a} - 2\sqrt{3a} =$

28) $-15\sqrt{17ab} - 10\sqrt{68ab} =$

29) $\sqrt{92x^2y} + x\sqrt{23y} =$

30) $5\sqrt{5a} + 4\sqrt{80a} =$

Answers of Worksheets – Chapter 5

Square Roots

1) 8
2) 2
3) 17
4) 0.5
5) 0.1
6) 0.3
7) 40
8) 1.5
9) 0
10) 0.2
11) 0.6
12) 0.9
13) 0.7
14) 1.1
15) 1.3
16) 0.4
17) 23
18) 25
19) 0.9
20) $2\sqrt{5}$
21) $5\sqrt{2}$
22) 26
23) $3\sqrt{30}$
24) $4\sqrt{2}$
25) 8
26) 56
27) 4
28) 17
29) 13
30) 15
31) $2\sqrt{19}$
32) 2
33) $6\sqrt{7}$
34) 420
35) 90
36) $6\sqrt{3}$

Simplifying radical expressions

1) $y\sqrt{13}$
2) $2x\sqrt{15x}$
3) $3\sqrt[3]{a}$
4) $9x$
5) $5\sqrt{6a}$
6) $3w\sqrt[3]{5}$
7) $10\sqrt{2x}$
8) $8\sqrt{3v}$
9) $4\sqrt[3]{x}$
10) $2x\sqrt{21x}$
11) $11x$
12) $2\sqrt[3]{6a}$
13) $4\sqrt{30}$
14) $15p\sqrt{7}$
15) $6m^3\sqrt{3}$
16) $3x.y\sqrt{22x}$
17) $13xy\sqrt{y}$
18) $5a^3$
19) $5xy\sqrt{2y}$
20) $8y$
21) $24x$
22) $60x$
23) $3y^3\sqrt[3]{7xy}$
24) $11xy\sqrt[3]{y^2}$
25) $15\sqrt{6a}$
26) $9\sqrt[3]{y}$
27) $9r\sqrt{2xyr}$
28) $90xz^3\sqrt{y}$
29) $15x\sqrt[3]{y^2}$
30) $14ac^2\sqrt{b}$
31) $40x^3y^{15}$

Multiplying radical expressions

1) 11
2) $5\sqrt{3}$
3) 6
4) $10\sqrt{5}$
5) $-10\sqrt{7}$
6) 12
7) 48
8) $-5\sqrt{15}$
9) $8\sqrt{55}$
10) $120\sqrt{21}$
11) $8\sqrt{2}+8$
12) $13x\sqrt{x}$

13) -42
14) $95x^3$
15) $5x\sqrt{3x}$
16) $-16x^3\sqrt{3}$
17) $-300x\sqrt{x}$
18) $-27\sqrt{3} - 27\sqrt{5}$
19) $6\sqrt{2x} - 4x$
20) $50x + 2\sqrt{50x}$
21) $18\sqrt{2r} + 6\sqrt{3r}$
22) $-108x^2\sqrt{5}$
23) $-216x$
24) $47v\sqrt{6v}$
25) -1
26) $16\sqrt{5} - 36$
27) 0
28) $49 - 19\sqrt{5}$
29) $28 - 3x$
30) $9r - 24\sqrt{3r} + 36$
31) $5\sqrt{n} - 64$
32) $-9 + 6\sqrt{2x} + 9\sqrt{2} - 12\sqrt{x}$

Simplifying radical expressions involving fractions

1) $\frac{\sqrt{6}}{2}$
2) $\frac{\sqrt{15}}{5}$
3) $\frac{\sqrt{2}}{2}$
4) $\frac{21\sqrt{5}}{5}$
5) $\frac{30\sqrt{2mr}}{m^3}$
6) $\frac{8\sqrt{2m}}{m}$
7) $\sqrt{15n}$
8) $\frac{2y\sqrt{xy}}{x}$
9) $-2(2 - \sqrt{2})$
10) $\frac{\sqrt{6x}\,(1-6\sqrt{x})}{6x}$
11) $\frac{2\sqrt{x}\,(\sqrt{x}+\sqrt{y})}{x-y}$
12) $\frac{15 + 3\sqrt{3} - 5\sqrt{5} - \sqrt{15}}{22}$
13) $\frac{31+15\sqrt{3}}{22}$
14) $-\frac{4(\sqrt{3}-1)}{3}$
15) $-\frac{5(2-\sqrt{15})}{11}$
16) -1
17) -3
18) $\frac{3-\sqrt{6}}{3}$
19) $-\frac{3\sqrt{11}+11+9\sqrt{5}+3\sqrt{55}}{2}$
20) $\frac{13\sqrt{7}+7+13\sqrt{13}+\sqrt{91}}{162}$
21) $5a^3\sqrt{b}$
22) $16\sqrt{6}\,m$

Adding and subtracting radical expressions

1) $2\sqrt{5}$
2) $21\sqrt{11}$
3) $-5\sqrt{3}$
4) $24 - 5\sqrt{3}$
5) $18\sqrt{5}$
6) $-14\sqrt{2}$
7) -120
8) $54\sqrt{2}$
9) -24
10) $26\sqrt{7}$
11) $30\sqrt{13}$
12) $-11\sqrt{5}$
13) $-22\sqrt{11}$
14) $2\sqrt{3}$
15) $19\sqrt{5}$
16) $-14\sqrt{7}$
17) 1
18) $16\sqrt{2}$
19) $30\sqrt{7}$
20) $5\sqrt{15}$
21) $15\sqrt{6}$
22) $36\sqrt{3}$
23) $-25\sqrt{5}$
24) 0
25) $4y\sqrt{14}$
26) $4n\sqrt{11m}$
27) $-34\sqrt{3a}$
28) $-35\sqrt{17ab}$
29) $-x\sqrt{23y}$
30) $21\sqrt{5a}$

Chapter 6:
Algebraic Expressions

Topics that you will practice in this chapter:

- ✓ Simplifying Variable Expressions
- ✓ Simplifying Polynomial Expressions
- ✓ Translate Phrases into an Algebraic Statement
- ✓ The Distributive Property
- ✓ Evaluating One Variable Expressions
- ✓ Evaluating Two Variables Expressions
- ✓ Combining like Terms

Mathematics is, as it were, a sensuous logic, and relates to philosophy as do the arts, music, and plastic art to poetry. — K. Shegel

Simplifying Variable Expressions

👁 **Simplify each expression.**

1) $3(x + 8) =$

2) $(-4)(7x - 3) =$

3) $11x + 8 - 7x =$

4) $-6 - 2x^2 - 9x^2 =$

5) $8 + 17x^2 + 6 =$

6) $9x^2 + 13x + 19x^2 =$

7) $7x^2 - 15x^2 + 3x =$

8) $8x^2 - 11x - 3x =$

9) $3x + 9(1 - 4x) =$

10) $14x + 2(20x - 4) =$

11) $6(-3x - 7) - 22 =$

12) $7x^2 + (-12x) =$

13) $x - 8 + 15 - 7x =$

14) $3 - 6x + 12 - 3x =$

15) $20x - 14 + 27 + 12x =$

16) $(-7)(6x - 5) + 14x =$

17) $11x - 4(3 - 7x) =$

18) $22x + 3(5x + 2) + 11 =$

19) $4(-3x + 8) + 10x =$

20) $16x - 3x(2x + 7) =$

21) $9x + 12x(2 - 4x) =$

22) $5x(-4x + 11) + 18x =$

23) $20x + 24x + 3x^2 =$

24) $7x(x - 7) - 28 =$

25) $7x - 12 + 5x + 3x^2 =$

26) $4x^2 - 9x - 12x =$

27) $8x - 22x^2 - 21x^2 - 11 =$

28) $9 + 3x^2 - 8x^2 - 27x =$

29) $14x + 2x^2 + 4x + 28 =$

30) $7x^2 + 45x + 12x^2 =$

31) $25 + 15x^2 + 9x - 3x^2 =$

32) $17x - 32x - 2x^2 + 30 =$

Simplifying Polynomial Expressions

✎ Simplify each polynomial.

1) $(5x^4 + 2x^2) - (11x + 6x^2) =$ _____

2) $(x^7 + 6x^4) - (8x^4 + 4x^2) =$ _____

3) $(24x^5 + 8x^3) - (x^3 - 12x^5) =$ _____

4) $14x - 9x^5 - 4(7x^5 + 7x^3) =$ _____

5) $(7x^4 - 5) + 2(4x^2 - 8x^4) =$ _____

6) $(9x^5 - 3x) - 3(8x^5 - 3x^4) =$ _____

7) $4(2x - 4x^4) - 5(3x^4 + x^2) =$ _____

8) $(4x^2 - 2x) - (5x^3 + 9x^2) =$ _____

9) $8x^4 - (9x^6 + 2x) + 2x^2 =$ _____

10) $x^5 - 3(x^3 + 2x) + 9x =$ _____

11) $(4x^2 - 2x^5) - (4x^5 - 2x^2) =$ _____

12) $8x^3 - 8x^5 + 17x^4 - 12x^5 =$ _____

13) $4x^3 - 9x^7 + 18x^7 - 24x^6 =$ _____

14) $4x^5 + 13x^3 - 17x^5 + 24x =$ _____

15) $7x^6 - 9x^7 + 5x^6 - 12x^3 =$ _____

16) $4x^4 + 19x - 3x^3 - 21x^4 =$ _____

Translate Phrases into an Algebraic Statement

✍ **Write an algebraic expression for each phrase.**

1) 13 multiplied by x. _____

2) Subtract 15 from y. _____

3) 22 divided by x. _____

4) 27 decreased by y. _____

5) Add y to 31. _____

6) The square of 7. _____

7) x raised to the seventh power. _____

8) The sum of five and a number. _____

9) The difference between forty-nine and y. _____

10) The quotient of eight and a number. _____

11) The quotient of the square of x and 34. _____

12) The difference between x and 14 is 41. _____

13) 7 times b reduced by the square of a. _____

14) Subtract the product of a and b from 51. _____

The Distributive Property

🖎 Use the distributive property to simply each expression.

1) $4(2 + 5x) =$

2) $5(2 + 4x) =$

3) $6(5x - 5) =$

4) $(6x - 3)(-7) =$

5) $(-4)(x + 8) =$

6) $(4 + 4x)6 =$

7) $(-5)(8 - 7x) =$

8) $-(-3 - 12x) =$

9) $(-8x + 3)(-5) =$

10) $(-5)(x - 11) =$

11) $-(8 - 2x) =$

12) $3(7 + 4x) =$

13) $4(8 + 3x) =$

14) $(-8x + 2)5 =$

15) $(4 - 7x)(-9) =$

16) $(-12)(3x + 5) =$

17) $(9 - 3x)5 =$

18) $4(4 + 7x) =$

19) $12(3x - 6) =$

20) $(-7x + 5)4 =$

21) $(4 - 9x)(-2) =$

22) $(-15)(2x - 3) =$

23) $(14 - 3x)3 =$

24) $(-5)(10x - 4) =$

25) $(5 - 7x)(-12) =$

26) $(-8)(2x + 9) =$

27) $(-5 + 8x)(-7) =$

28) $(-6)(2 - 15x) =$

29) $13(4x - 6) =$

30) $(-15x + 13)(-4) =$

31) $(-9)(3x - 2) + 2(x + 5) =$

32) $(-9)(2x + 2) - (7 + 4x) =$

Evaluating One Variable Expressions

✍ **Evaluate each expression using the value given.**

1) $8 - x, x = 5$

2) $x - 10, x = 6$

3) $3x - 6, x = 5$

4) $x - 15, x = -2$

5) $12 - x, x = 4$

6) $x + 7, x = 1$

7) $2x + 9, x = 7$

8) $x + (-4), x = -7$

9) $2x + 9, x = 4$

10) $3x + 10, x = -2$

11) $18 + 2x - 4, x = -1$

12) $18 - 6x, x = 2$

13) $8x - 2, x = 4$

14) $2x - 17, x = 8$

15) $13x - 12, x = 3$

16) $8 - 5x, x = -2$

17) $3(5x + 4), x = 5$

18) $4(-2x - 7), x = 3$

19) $7x - 5x + 12, x = 2$

20) $(8x + 4) \div 2, x = 6$

21) $(x + 15) \div 4, x = 9$

22) $6x - 10 + 3x, x = -5$

23) $(7 - 4x)(-3), x = -3$

24) $12x^2 + 5x - 4, x = 2$

25) $x^2 - 15x, x = -4$

26) $3x(3 - 6x), x = 2$

27) $13x + 8 - 6x^2, x = -3$

28) $(-2)(15x - 11 + 4x), x = 4$

29) $(-6) + \frac{x}{6} + x, x = 18$

30) $(-9) + \frac{x}{4}, x = 32$

31) $\left(-\frac{45}{x}\right) - 5 + 2x, x = 9$

32) $\left(-\frac{36}{x}\right) - 9 + 3x, x = 3$

Evaluating Two Variables Expressions

✎ **Evaluate each expression using the values given.**

1) $5x - y$,
   $x = 4, y = 3$

2) $3x + 2y$,
   $x = -2, y = 2$

3) $-6a + 5b$,
   $a = 3, b = 1$

4) $3x + 7 - y$,
   $x = 8, y = 4$

5) $5z + 12 - 3k$,
   $z = 5, k = 2$

6) $6(-x - 3y)$,
   $x = 5, y = 4$

7) $7a + 4b$,
   $a = 3, b = 5$

8) $8x \div 4y$,
   $x = 6, y = 4$

9) $2x + 18 + 4y$,
   $x = -3, y = 3$

10) $5a - (18 - 2b)$,
    $a = 5, b = 8$

11) $6z + 12 + 3k$,
    $z = -3, k = 3$

12) $2xy + 6 + 7x$,
    $x = 5, y = 3$

13) $6x + 2y - 9 + 3$,
    $x = 3, y = 2$

14) $\left(-\frac{21}{x}\right) + 6 + 3y$,
    $x = 7, y = 4$

15) $(-4)(-3a - b)$,
    $a = 2, b = 6$

16) $18 + 4x + 9 - 5y$,
    $x = 6, y = 4$

17) $7x + 5 - 6y + 11$,
    $x = 9, y = 3$

18) $9 + 4(-5x - 3y)$,
    $x = 4, y = 5$

19) $3x + 15 + 6y$,
    $x = 2, y = 4$

20) $7a - (4a - 2b) + 8$,
    $a = 3, b = 1$

Combining like Terms

✏️ **Simplify each expression.**

1) $7x + 2x + 8 =$

2) $3(6x - 2) =$

3) $10x - 12x + 8 =$

4) $20x - 32x + 14 =$

5) $16x - 6x - 12 =$

6) $18x - 21 + 4x =$

7) $15 - (3x + 9) =$

8) $-14x + 7 - 11x =$

9) $5x - 10 - 3x + 1 =$

10) $24x + 7x - 22 =$

11) $14x + 8x - 2 =$

12) $(-4x + 2)8 =$

13) $34 + 6x + 8x - 4 =$

14) $3(x - 8x) - 5 =$

15) $4(2x + 7) + 5x =$

16) $x - 27 - 9x =$

17) $3(5 + 4x) - 8x =$

18) $41x + 24 + 3x =$

19) $(-8x) + 30 + 15x =$

20) $(-4x) - 12 + 19x =$

21) $5(2x + 6) + 9x =$

22) $3(6 - 7x) - 11x =$

23) $-8x - (16 - 14x) =$

24) $(-9) - (6)(5x + 9) =$

25) $(-4)(6x - 5) - 12x =$

26) $-34x + 14 + 9x - 21x =$

27) $5(-13x + 6) - 24x =$

28) $-7x - 20 + 15x =$

29) $42x - 31x + 15 - 12x =$

30) $4(8x + 5x) - 17 =$

31) $54 - 22x - 28 - 19x =$

32) $-9(-7x - 11x) + 58x =$

Answers of Worksheets – Chapter 6

Simplifying Variable Expressions

1) $3x + 24$
2) $-28x + 12$
3) $4x + 8$
4) $-11x^2 - 6$
5) $17x^2 + 14$
6) $28x^2 + 13x$
7) $-8x^2 + 3x$
8) $8x^2 - 14x$
9) $-33x + 9$
10) $54x - 8$
11) $-18x - 64$
12) $7x^2 - 12x$
13) $-6x + 7$
14) $-9x + 15$
15) $32x + 13$
16) $-28x + 35$
17) $39x - 12$
18) $37x + 17$
19) $-2x + 32$
20) $-6x^2 - 5x$
21) $-48x^2 + 33x$
22) $-20x^2 + 73x$
23) $3x^2 + 44x$
24) $7x^2 - 49x - 28$
25) $3x^2 + 12x - 12$
26) $4x^2 - 21x$
27) $-43x^2 + 8x - 11$
28) $-5x^2 - 27x + 9$
29) $2x^2 + 18x + 28$
30) $19x^2 + 45x$
31) $12x^2 + 9x + 25$
32) $-2x^2 - 15x + 30$

Simplifying Polynomial Expressions

1) $5x^4 - 4x^2 - 11x$
2) $x^7 - 2x^4 - 4x^2$
3) $36x^5 + 7x^3$
4) $-37x^5 - 28x^2 + 14x$
5) $-9x^4 + 8x^2 - 5$
6) $-15x^5 + 9x^4 - 3x$
7) $-31x^3 - 5x^2 + 8x$
8) $-5x^3 - 5x^2 - 2x$
9) $-9x^6 + 8x^4 + 2x^2 - 2x$
10) $x^5 - 3x^3 + 3x$
11) $-6x^5 + 6x^2$
12) $-20x^5 + 17x^4 + 8x^3$
13) $9x^7 - 24x^6 + 4x^3$
14) $-13x^5 + 13x^3 + 24x$
15) $-9x^7 + 12x^6 - 12x^3$
16) $-17x^4 - 3x^3 + 19x$

Translate Phrases into an Algebraic Statement

1) $13x$
2) $y - 15$
3) $\frac{22}{x}$
4) $27 - y$
5) $y + 31$
6) 7^2
7) x^7
8) $5 + x$
9) $49 - y$
10) $\frac{8}{x}$
11) $\frac{x^2}{34}$
12) $x - 14 = 41$
13) $7b - a^2$
14) $51 - ab$

The Distributive Property

1) $20x + 8$
2) $20x + 10$
3) $30x - 30$
4) $-42x + 21$
5) $-4x - 32$
6) $24x + 24$
7) $35x - 40$
8) $12x + 3$

9) $40x - 15$
10) $-5x + 55$
11) $2x - 8$
12) $12x + 21$
13) $12x + 32$
14) $-40x + 10$

15) $63x - 36$
16) $-36x - 60$
17) $-15x + 45$
18) $28x + 16$
19) $36x - 72$
20) $-28x + 20$

21) $18x - 8$
22) $-30x + 45$
23) $-9x + 42$
24) $-50x + 20$
25) $84x - 60$
26) $-16x - 72$

27) $56x - 35$
28) $90x - 12$
29) $52x - 78$
30) $60x - 52$
31) $-25x + 28$
32) $-22x - 25$

Evaluating One Variables

1) 3
2) -4
3) 9
4) -17
5) 8
6) 8
7) 23
8) -11

9) 17
10) 4
11) 12
12) 6
13) 30
14) -1
15) 27
16) 18

17) 87
18) -52
19) 16
20) 26
21) 6
22) -55
23) -57
24) 54

25) 76
26) -54
27) -85
28) -130
29) 15
30) -1
31) 8
32) -12

Evaluating Two Variables

1) 17
2) -2
3) -13
4) 27
5) 31

6) -102
7) 41
8) 3
9) 24
10) 23

11) 3
12) 71
13) 16
14) 15
15) 48

16) 31
17) 61
18) -131
19) 45
20) 19

Combining like Terms

1) $9x + 8$
2) $18x - 6$
3) $-2x + 8$
4) $-12x + 14$
5) $10x - 12$
6) $22x - 22$
7) $-3x + 6$
8) $-25x + 7$

9) $2x - 9$
10) $31x - 22$
11) $22x - 2$
12) $-32x + 16$
13) $14x + 30$
14) $-21x - 5$
15) $13x + 28$
16) $-8x - 27$

17) $4x + 15$
18) $44x + 24$
19) $7x + 30$
20) $15x - 12$
21) $19x + 30$
22) $-32x + 18$
23) $6x - 16$
24) $-30x - 63$

25) $-36x + 20$
26) $-46x + 14$
27) $-89x + 30$
28) $8x - 20$
29) $-x + 15$
30) $52x - 17$
31) $-41x + 26$
32) $220x$

Chapter 7:
Equations and Inequalities

Topics that you will practice in this chapter:

- ✓ One–Step Equations
- ✓ Multi–Step Equations
- ✓ Graphing Single–Variable Inequalities
- ✓ One–Step Inequalities
- ✓ Multi-Step Inequalities
- ✓ Systems of Equations
- ✓ Systems of Equations Word Problems
- ✓ Finding Midpoint
- ✓ Finding Distance of Two Points

"Life is a math equation. In order to gain the most, you have to know how to convert negatives into positives." – Anonymous

One–Step Equations

✍ **Find the answer for each equation.**

1) $3x = 90, x =$ ___

2) $5x = 35, x =$ ___

3) $9x = 36, x =$ ___

4) $25x = 150, x =$ ___

5) $x + 18 = 23, x =$ ___

6) $x - 3 = 8, x =$ ___

7) $x - 7 = 4, x =$ ___

8) $x + 22 = 30, x =$ ___

9) $x - 11 = 6, x =$ ___

10) $24 = 28 + x, x =$ ___

11) $x - 5 = 7, x =$ ___

12) $9 - x = -7, x =$ ___

13) $43 = -8 + x, x =$ ___

14) $x - 23 = -38, x =$ ___

15) $x + 45 = -27, x =$ ___

16) $42 = 56 - x, x =$ ___

17) $-18 + x = -32, x =$ ___

18) $x - 13 = 7, x =$ ___

19) $35 = x - 10, x =$ ___

20) $x - 8 = -21, x =$ ___

21) $x - 54 = -20, x =$ ___

22) $x - 42 = -47, x =$ ___

23) $x - 8 = 29, x =$ ___

24) $-93 = x - 51, x =$ ___

25) $x + 15 = 37, x =$ ___

26) $108 = 12x, x =$ ___

27) $x - 33 = 27, x =$ ___

28) $x - 12 = 23, x =$ ___

29) $72 - x = 18, x =$ ___

30) $x + 34 = 58, x =$ ___

31) $21 - x = -9, x =$ ___

32) $x - 59 = -80, x =$ ___

Multi-Step Equations

✎ **Find the answer for each equation.**

1) $3x + 1 = 7$

2) $-x + 10 = 9$

3) $5x - 13 = 7$

4) $-(4 - x) = 5$

5) $3x - 8 = 16$

6) $15x - 13 = 17$

7) $3x - 28 = 2$

8) $9x + 21 = 39$

9) $14x + 17 = 45$

10) $-14(8 + x) = 70$

11) $8(10 + x) = 32$

12) $16 = -(x - 8)$

13) $5(7 - 3x) = 50$

14) $-19 = -(3x + 7)$

15) $30(3 + x) = 60$

16) $9(x - 12) = 54$

17) $-24 = 3x + 5x$

18) $5x + 28 = -2x - 7$

19) $9(5 + 4x) = -99$

20) $18 - x = -12 - 6x$

21) $4 - 4x = 28 - 2x$

22) $15 + 12x = -15 + 8x$

23) $54 = (-3x) - 8 + 8$

24) $12 = 7x - 18 + 5x$

25) $-18 = -9x - 42 + 5x$

26) $11x - 6 = -33 + 8x$

27) $8x - 42 = 3x + 3$

28) $-15 - 8x = 4(5 - x)$

29) $x - 9 = -5(9 - 2x)$

30) $14x - 65 = -x - 110$

31) $3x - 129 = -3(11 + 7x)$

32) $-7x - 20 = 2x + 43$

Graphing Single–Variable Inequalities

✎ **Draw a graph for each inequality.**

1) $x \leq 7$

2) $x \leq -1.5$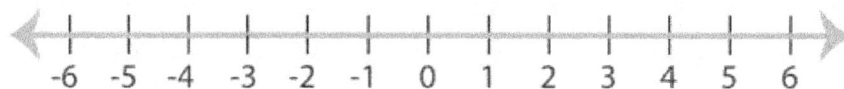

3) $x < -4$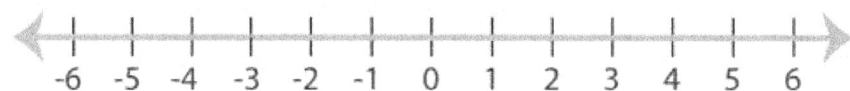

4) $x > 2.5$

5) $x > 1.3$

6) $x < 4$

7) $x < 2.4$

8) $x > -\frac{18}{10}$

One–Step Inequalities

✏ **Find the answer for each inequality and graph it.**

1) $x + 3 > -5$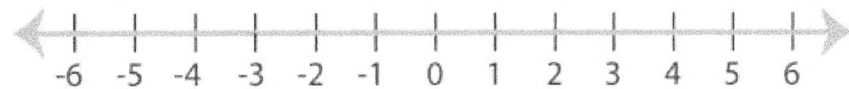

2) $x - 4 < 1$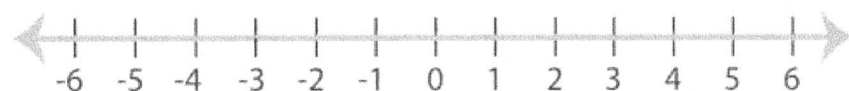

3) $7x < 42$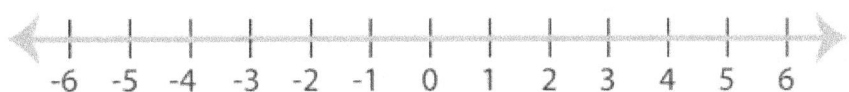

4) $13 + x > 12$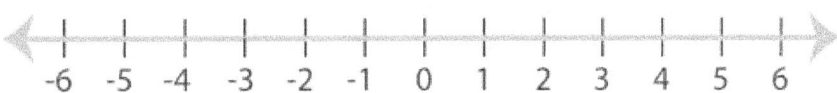

5) $x + 20 < 13$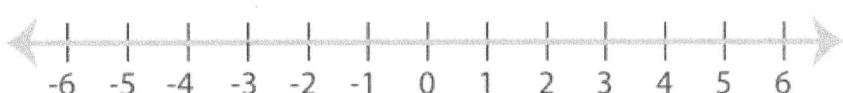

6) $14x \leq 42$

7) $11x \leq -44$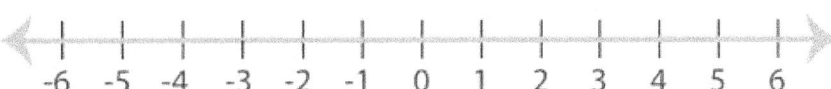

8) $x + 26 > 35$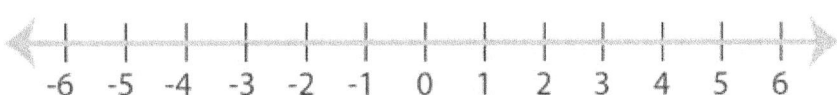

Multi-Step Inequalities

✏️ **Calculate each inequality.**

1) $x - 8 \leq 12$

2) $9 - 3x \leq 18$

3) $4x - 7 \leq 9$

4) $8x - 9 \geq 15$

5) $x - 19 \geq 24$

6) $5x - 15 \leq 40$

7) $7x - 4 \leq 24$

8) $-18 + 8x \leq 22$

9) $9(x - 8) \leq 27$

10) $4x - 8 \leq 16$

11) $11x - 42 < 22$

12) $10x - 18 < 52$

13) $17 - 9x \geq -46$

14) $32 + 2x < 68$

15) $8 + 8x \geq 80$

16) $11 + 6x < 65$

17) $9x - 13 < 23$

18) $8(12 - 4x) \geq -68$

19) $-(2 + 5x) < 42$

20) $14 - 9x \geq -31$

21) $-5(x - 3) > 65$

22) $\dfrac{2x + 8}{3} \leq 12$

23) $\dfrac{8x + 16}{4} \leq 24$

24) $\dfrac{2x - 22}{9} > 8$

25) $7 + \dfrac{x}{4} < 21$

26) $\dfrac{32x}{16} - 4 < 6$

27) $\dfrac{12x + 36}{22} > 3$

28) $42 + \dfrac{x}{3} < 15$

Systems of Equations

✏ **Calculate each system of equations.**

1) $-6x + 7y = 8$ $x = \_\_\_$
   $x + 4y = 9$ $y = \_\_\_$

2) $-4x + 12y = 12$ $x = \_\_\_$
   $14x - 16y = 10$ $y = \_\_\_$

3) $y = -9$ $x = \_\_\_$
   $2x - 5y = 12$ $y = \_\_\_$

4) $4y = -4x + 20$ $x = \_\_\_$
   $8x - 2y = -12$ $y = \_\_\_$

5) $10x - 9y = -13$ $x = \_\_\_$
   $-5x + 3y = 11$ $y = \_\_\_$

6) $-6x - 8y = 10$ $x = \_\_\_$
   $4x - 8y = 20$ $y = \_\_\_$

7) $5x - 14y = -23$ $x = \_\_\_$
   $-6x + 7y = 8$ $y = \_\_\_$

8) $-4x + 3y = 3$ $x = \_\_\_$
   $-x + 2y = 5$ $y = \_\_\_$

9) $-4x + 5y = 15$ $x = \_\_\_$
   $-3x + 4y = -10$ $y = \_\_\_$

10) $-6x - 6y = -21$ $x = \_\_\_$
    $-6x + 6y = -66$ $y = \_\_\_$

11) $12x - 21y = 6$ $x = \_\_\_$
    $-6x - 3y = -12$ $y = \_\_\_$

12) $-4x - 4y = -14$ $x = \_\_\_$
    $4x - 4y = 44$ $y = \_\_\_$

13) $4x + 5y = 3$ $x = \_\_\_$
    $3x - y = 6$ $y = \_\_\_$

14) $3x - 2y = 2$ $x = \_\_\_$
    $10x - 10y = 20$ $y = \_\_\_$

15) $5x + 8y = 14$ $x = \_\_\_$
    $-3x - 2y = -3$ $y = \_\_\_$

16) $8x + 5y = 4$ $x = \_\_\_$
    $-3x - 4y = 15$ $y = \_\_\_$

Systems of Equations Word Problems

✎ **Find the answer for each word problem.**

1) Tickets to a movie cost $6 for adults and $4 for students. A group of friends purchased 9 tickets for $50.00. How many adults ticket did they buy? ____

2) At a store, Eva bought two shirts and five hats for $77.00. Nicole bought three same shirts and four same hats for $84.00. What is the price of each shirt? ____

3) A farmhouse shelters 10 animals, some are pigs, and some are ducks. Altogether there are 36 legs. How many pigs are there? ____

4) A class of 85 students went on a field trip. They took 24 vehicles, some cars and some buses. If each car holds 3 students and each bus hold 16 students, how many buses did they take? ____

5) A theater is selling tickets for a performance. Mr. Smith purchased 8 senior tickets and 10 child tickets for $248 for his friends and family. Mr. Jackson purchased 4 senior tickets and 6 child tickets for $132. What is the price of a senior ticket? $____

6) The difference of two numbers is 15. Their sum is 33. What is the bigger number? $____

7) The sum of the digits of a certain two-digit number is 7. Reversing its digits increase the number by 9. What is the number? ____

8) The difference of two numbers is 11. Their sum is 25. What are the numbers? _____

9) The length of a rectangle is 5 meters greater than 2 times the width. The perimeter of rectangle is 28 meters. What is the length of the rectangle? _____

10) Jim has 23 nickels and dimes totaling $2.40. How many nickels does he have? ____

Finding Midpoint

✎ **Find the midpoint of the line segment with the given endpoints.**

1) $(-4, -6), (2, 4)$

2) $(13, 5), (-1, 5)$

3) $(11, -4), (3, 14)$

4) $(-15, -6), (3, 9)$

5) $(7, -8), (13, -12)$

6) $(-14, -8), (8, -12)$

7) $(9, 2), (-9, 22)$

8) $(-8, 10), (-8, 4)$

9) $(-7, 7), (23, -15)$

10) $(3, 17), (19, -5)$

11) $(-4, 13), (7, 9)$

12) $(11, 8), (-3, -6)$

13) $(-4, 12), (0, 6)$

14) $(34, 12), (18, -28)$

15) $(15, 6), (-1, 0)$

16) $(-11, -13), (-13, 19)$

17) $(12, 4), (8, 16)$

18) $(-2, -7), (18, -21)$

19) $(18, 13), (-6, 5)$

20) $(10, -4), (0, 18)$

21) $(4, -4), (8, -20)$

22) $(25, 5), (-11, -17)$

23) $(8, 12), (16, -2)$

24) $(14, -20), (8, 14)$

✎ **Find the answer for each problem.**

25) One endpoint of a line segment is $(6, 8)$ and the midpoint of the line segment is $(1, 6)$. What is the other endpoint? _____

26) One endpoint of a line segment is $(-7, 5)$ and the midpoint of the line segment is $(1, 3)$. What is the other endpoint? _____

27) One endpoint of a line segment is $(-6, -10)$ and the midpoint of the line segment is $(2, 9)$. What is the other endpoint? _____

Finding Distance of Two Points

✎ **Find the distance between each pair of points.**

1) $(5, 9), (-11, -3)$

2) $(-6, 2), (-2, 6)$

3) $(-8, -1), (-3, 8)$

4) $(-8, -2), (2, 22)$

5) $(6, -4), (-12, -28)$

6) $(-6, 0), (-2, 3)$

7) $(8, 12), (8, 6)$

8) $(12, -10), (12, -2)$

9) $(15, 27), (-33, -9)$

10) $(10, -2), (6, -14)$

11) $(1, 0), (6, 12)$

12) $(8, 4), (3, -8)$

13) $(3, 2), (-5, -11)$

14) $(-10, 12), (6, 42)$

15) $(0, 16), (-8, 10)$

16) $(5, 0), (30, 60)$

17) $(3, 5), (-5, -10)$

18) $(-4, 6), (4, 3)$

19) $(7, 2), (-8, -18)$

20) $(-10, 8), (14, 18)$

✎ **Find the answer for each problem.**

21) Triangle ABC is a right triangle on the coordinate system and its vertices are $(-5, 7), (-5, 1),$ and $(1, 1)$. What is the area of triangle ABC? _____

22) Three vertices of a triangle on a coordinate system are $(1, 1), (7, 1),$ and $(1, 9)$. What is the perimeter of the triangle? _____

23) Four vertices of a rectangle on a coordinate system are $(-2, 4), (-2, 7), (4, 4),$ and $(4, 7)$. What is its perimeter? _____

Answers of Worksheets – Chapter 7

One–Step Equations

1) 30	9) 17	17) -14	25) 22
2) 7	10) -4	18) 20	26) 9
3) 4	11) 12	19) 45	27) 60
4) 6	12) 16	20) -13	28) 35
5) 5	13) 51	21) 34	29) 54
6) 11	14) -15	22) -5	30) 24
7) 11	15) -72	23) 37	31) 30
8) 8	16) 14	24) -42	32) -21

Multi–Step Equations

1) 2	9) 2	17) -3	25) -6
2) 1	10) -13	18) -5	26) -9
3) 4	11) -6	19) -4	27) 9
4) 9	12) -8	20) -6	28) -8.75
5) 8	13) -1	21) -12	29) 4
6) 2	14) 4	22) -7.5	30) -3
7) 10	15) -1	23) -18	31) 4
8) 2	16) 18	24) 2.5	32) -7

Graphing Single–Variable Inequalities

1)

2)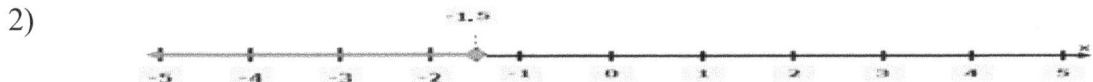

3)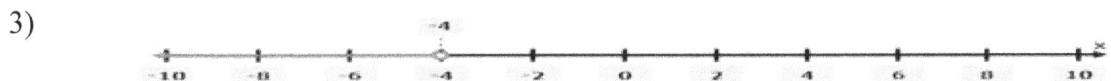

4)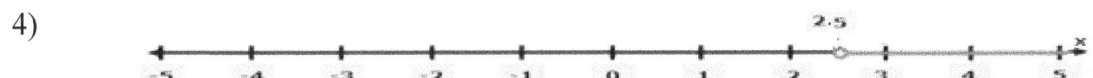

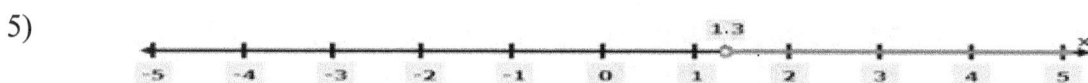

5) [number line with open circle at 1.3, arrow left]

6) [number line with open circle at 4, arrow right]

7) [number line with open circle at 2.4, arrow right]

8) [number line with open circle at −1.8, arrow right]

One–Step Inequalities

1) [number line with open circle at −8, arrow left]

2) [number line with closed circle at 5, arrow right]

3) [number line with closed circle at 6, arrow right]

4) [number line with open circle at −1, arrow left]

5) [number line with open circle at −7, arrow left]

6) [number line with closed circle at 3, arrow right]

7) [number line with closed circle at −4, arrow left]

8) [number line with open circle at 9, arrow right]

Multi-Step Inequalities

1) $x \leq 20$
2) $x \geq -3$
3) $x \leq 4$
4) $x \geq 3$
5) $x \geq 43$
6) $x \leq 11$
7) $x \leq 4$
8) $x \leq 5$
9) $x \leq 11$
10) $x \leq 6$
11) $x < 64/11$
12) $x < 7$
13) $x \leq 7$
14) $x < 18$
15) $x \geq 9$
16) $x < 9$

17) $x < 4$	20) $x \leq 5$	23) $x \leq 10$	26) $x < 9/4$
18) $x \leq 41/8$	21) $x < -10$	24) $x > 47$	27) $x > 2.5$
19) $x > -44/5$	22) $x \leq 14$	25) $x < 56$	28) $x < -81$

Systems of Equations

1) $x = 1, y = 2$
2) $x = 3, y = 2$
3) $x = -\frac{33}{2}$
4) $x = -\frac{1}{5}, y = \frac{26}{5}$
5) $x = -4, y = -3$
6) $x = 1, y = -2$
7) $x = 1, y = 2$
8) $x = \frac{9}{5}, y = \frac{17}{5}$
9) $x = -110, y = -85$
10) $x = -\frac{15}{4}, y = \frac{29}{4}$
11) $x = \frac{5}{3}, y = \frac{2}{3}$
12) $x = -\frac{15}{4}, y = \frac{29}{4}$
13) $x = \frac{33}{19}, y = -\frac{15}{19}$
14) $x = -2, y = -4$
15) $x = -\frac{2}{7}, y = \frac{27}{14}$
16) $x = \frac{91}{17}, y = -\frac{132}{17}$

Systems of Equations Word Problems

1) 7
2) $16
3) 8
4) 1
5) $21
6) 24
7) 43
8) 18, 7
9) 11 meters
10) 18

Finding Midpoint

1) $(-1, -1)$
2) $(6, 5)$
3) $(7, 5)$
4) $(-6, 1.5)$
5) $(10, -10)$
6) $(-3, -10)$
7) $(0, 12)$
8) $(-8, 7)$
9) $(8, -4)$
10) $(11, 6)$
11) $(1.5, 11)$
12) $(4, 1)$
13) $(-2, 9)$
14) $(26, -8)$
15) $(7, 3)$
16) $(-12, 3)$
17) $(10, 10)$
18) $(8, -14)$
19) $(6, 9)$
20) $(5, 7)$
21) $(6, -12)$
22) $(7, -6)$
23) $(12, 5)$
24) $(11, -3)$
25) $(-4, 4)$
26) $(9, 1)$
27) $(10, 28)$

Finding Distance of Two Points

1) 20
2) $4\sqrt{2}$
3) $\sqrt{106}$
4) 26
5) 30
6) 10
7) 6
8) 8
9) 60

10) $4\sqrt{10}$
11) 13
12) 13
13) $\sqrt{233}$
14) 34

15) 10
16) 65
17) 17
18) $\sqrt{73}$
19) 25

20) 26
21) 18 square units
22) 24 units
23) 18 units

Chapter 8:
Linear Functions

Topics that you will practice in this chapter:

- ✓ Relation and Function
- ✓ Finding Slope
- ✓ Graphing Lines Using Line Equation
- ✓ Writing Linear Equations
- ✓ Graphing Linear Inequalities
- ✓ Write an Equation from a Graph
- ✓ Finding Rate of Change, x–intercept and y–intercept
- ✓ Slope-Intercept Form
- ✓ Point-Slope Form
- ✓ Graphing Lines of Equations
- ✓ Equation of parallel or perpendicular lines
- ✓ Equations of horizontal and vertical lines
- ✓ Graphing Absolute Value Equation

"Sometimes the questions are complicated, and the answers are simple." – Dr. Seuss

Relation and Functions

State the domain and range of each relation. Then determine whether each relation is a function.

1)
Function:
....
Domain:
....
Range:
....

Mapping: 2→1, 4→0, 8→5, 9→7, 10→14 (with 2 also mapping to 5)

2)
Function:
....
Domain:
....
Range:
....

x	y
2	3
1	0
−1	−4
7	−4
9	5

3)
Function:
....
Domain:
....
Range:
....

(downward parabola graph)

4) $\{(6, -8), (3, -2), (2, 4), (3, 0), (5, 9)\}$
Function:
....
Domain:
....
Range:
....

5)
Function:
....
Domain:
....
Range:
....

(sideways parabola graph opening right)

6)
Function:
....
Domain:
....
Range:
....

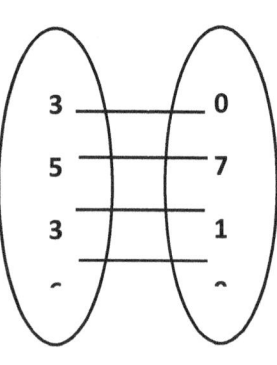

Mapping: 3→0, 5→7, 3→1

Finding Slope

🔖 **Find the slope of each line.**

1) $y = 2x + 5$

2) $y = -x + 17$

3) $y = 4x + 16$

4) $y = -3x + 15$

5) $y = 27 + 7x$

6) $y = 11 - 4x$

7) $y = 7x + 14$

8) $y = -8x + 18$

9) $y = -9x + 15$

10) $y = 8x - 13$

11) $y = \frac{1}{5}x + 9$

12) $y = -\frac{3}{7}x + 19$

13) $-3x + 6y = 17$

14) $4x + 4y = 16$

15) $8y - 3x = 32$

16) $11y - 3x = 42$

🔖 **Find the slope of the line through each pair of points.**

17) $(1, 8), (5, 16)$

18) $(-2, 14), (2, 18)$

19) $(7, -1), (3, 9)$

20) $(-4, -4), (2, 14)$

21) $(16, -1), (4, 11)$

22) $(-21, 5), (-10, 38)$

23) $(8, 11), (12, 19)$

24) $(22, -22), (10, 14)$

25) $(21, -15), (19, -13)$

26) $(11, 10), (7, -2)$

27) $(5, 4), (9, 16)$

28) $(34, -87), (22, 45)$

Graphing Lines Using Line Equation

✎ **Sketch the graph of each line.**

1) $y = x - 5$

2) $y = -3x + 4$

3) $x - 2y = 0$

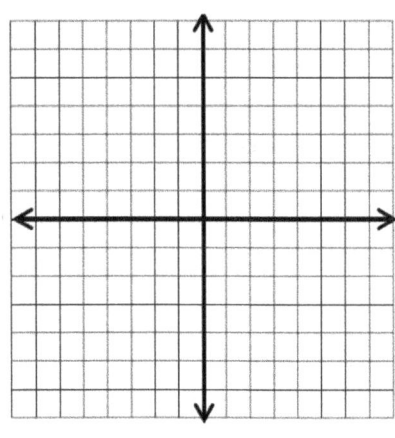 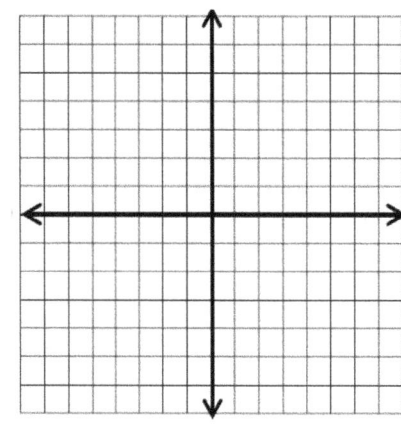 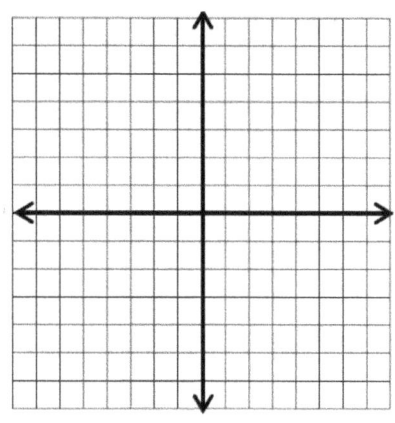

4) $x + y = -4$

5) $4x + 3y = -2$

6) $y - 3x + 6 = 0$

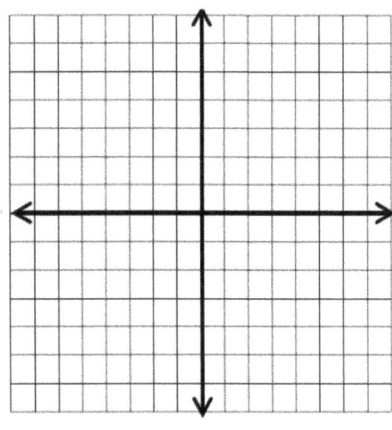 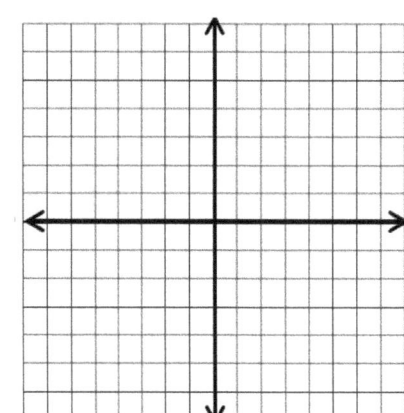 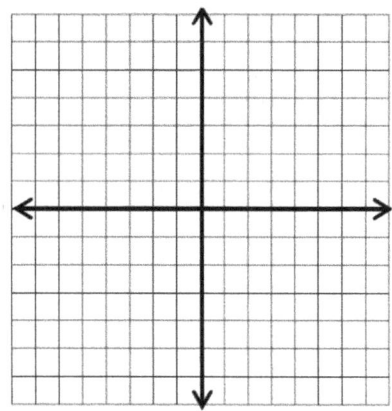

Writing Linear Equations

✎ **Write the equation of the line through the given points.**

1) Through: $(6, -10), (10, 14)$

2) Through: $(10, 4), (4, 22)$

3) Through: $(-6, 4), (2, 12)$

4) Through: $(15, 11), (3, -1)$

5) Through: $(-5, 33), (9, 5)$

6) Through: $(20, 5), (17, 2)$

7) Through: $(24, -4), (16, 4)$

8) Through: $(-18, 57), (33, -45)$

9) Through: $(10, 12), (8, 18)$

10) Through: $(25, 41), (33, -7)$

11) Through: $(-6, 9), (-8, -7)$

12) Through: $(8, 8), (4, -8)$

13) Through: $(6, -10), (10, 6)$

14) Through: $(10, -24), (-8, 12)$

15) Through: $(10, 10), (-2, -4)$

16) Through: $(-7, 35), (11, -31)$

✎ **Find the answer for each problem.**

17) What is the equation of a line with slope 3 and intercept 11? _____

18) What is the equation of a line with slope 5 and intercept 15? _____

19) What is the equation of a line with slope 7 and passes through point $(3, 2)$? _____

20) What is the equation of a line with slope -3 and passes through point $(-2, 5)$? _____

21) The slope of a line is -6 and it passes through point $(-2, 1)$. What is the equation of the line? _____

22) The slope of a line is 5 and it passes through point $(-4, 2)$. What is the equation of the line? _____

Graphing Linear Inequalities

✎ **Sketch the graph of each linear inequality.**

1) $y > 3x - 5$

2) $y < 2x + 1$

3) $y \leq -4x - 5$

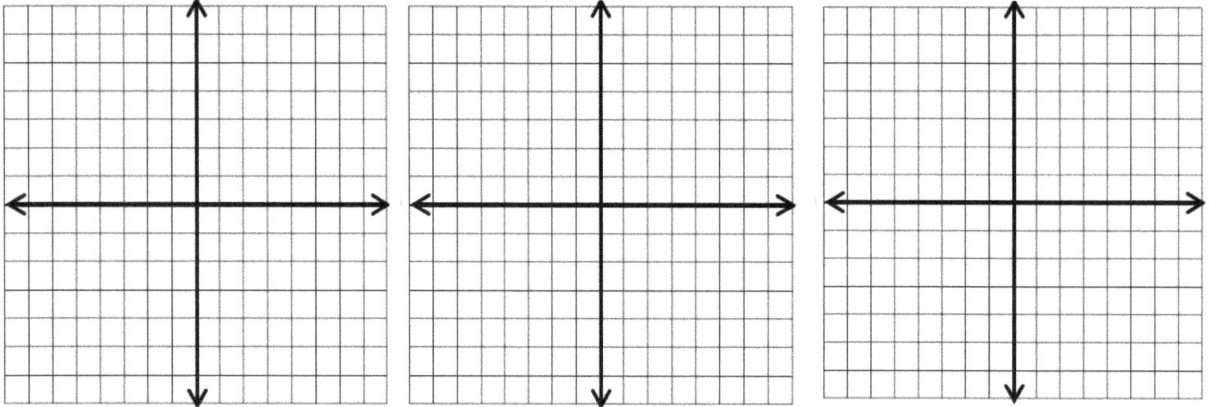

4) $2y \geq 12 + 4x$

5) $-5y < x - 15$

6) $3y \geq -9x + 6$

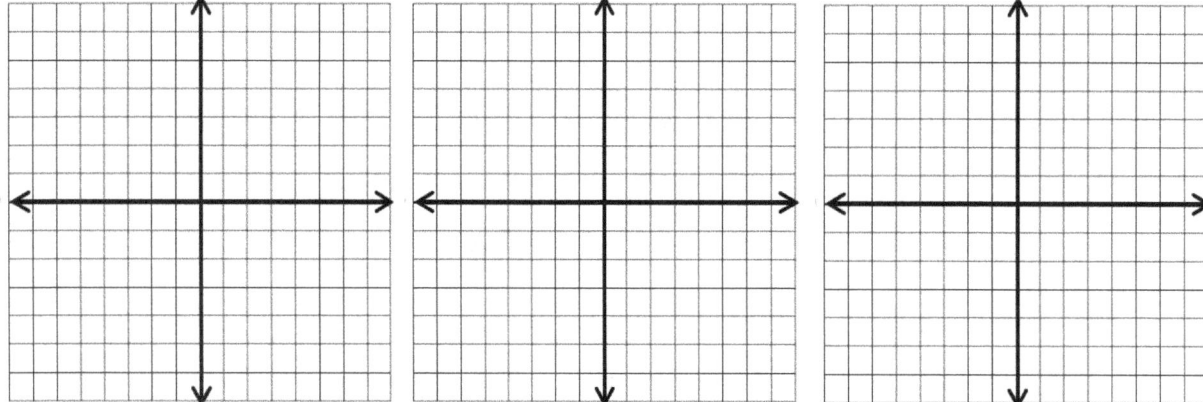

Write an Equation from a Graph

Write the slope intercept form of the equation of each line

1)

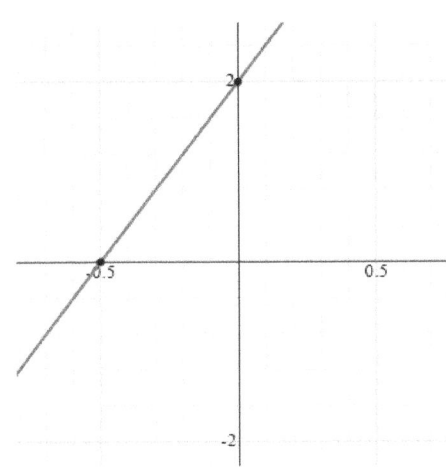

2)

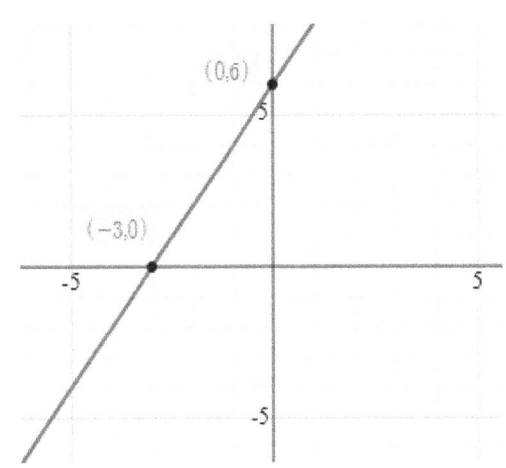

3)

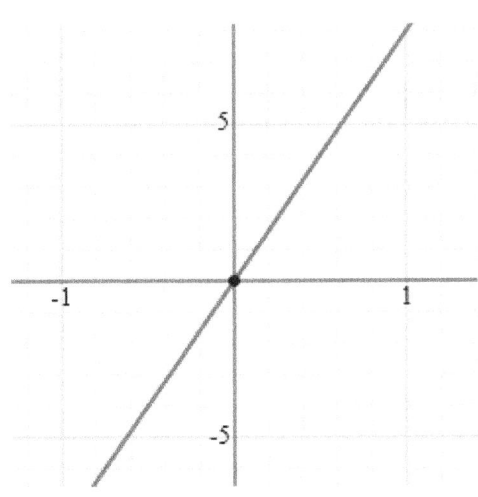

4)

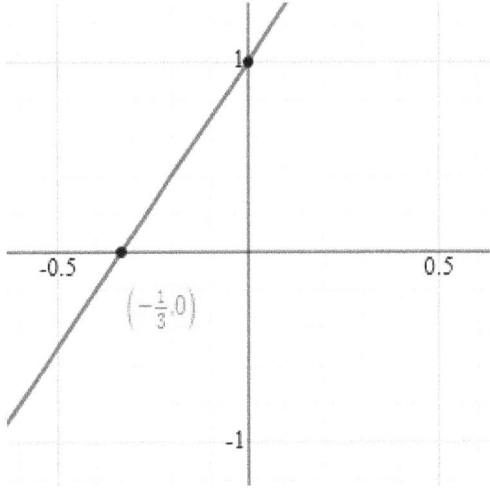

Rate of change

What is the average rate of change of the function?

1) $f(x) = 2x^2 + 3$ from $x = 2$ to $x = 5$?

2) $f(x) = -x^2 - 6$, from $x = 3$ to $x = 7$?

3) $f(x) = 2x^3 + 5$, from $x = 0$ to $x = 1$?

x and y intercepts

Find the x and y intercepts for the following equations.

1) $4x + 2y = 12$

2) $y = x + 4$

3) $3x = y + 18$

4) $x + y = -6$

5) $7x - 5y = 8$

6) $5y - 4x + 12 = 0$

7) $\frac{3}{5}x + \frac{1}{5}y + \frac{3}{4} = 0$

8) $6x - 24 = 0$

9) $28 - 7y = 0$

10) $-3x - 5y + 45 = 15$

Find the value of b: The line that passes through each pair of points has the given slope.

11) $(8, -3), (4, b), m = 2$

12) $(b, 5), (-5, 2), m = \frac{1}{2}$

13) $(-3, b), (3, 5), m = \frac{1}{3}$

14) $(-2, 2), (b, 9), m = 1\frac{3}{4}$

Slope–intercept Form

Write the slope–intercept form of the equation of each line.

1) $-15x + y = 7$

2) $-3(5x + y) = 36$

3) $-7x - 21y = -42$

4) $4x + 12 = -8y$

5) $2x - 5y = 15$

6) $14x - 10y = -20$

7) $27x - 9y = -54$

8) $6x - 5y + 36 = 0$

9) $-\frac{1}{4}y = -3x + 5$

10) $8 - 2y - 5x = 0$

11) $-2y = -3x - 8$

12) $12x + 7y = -21$

13) $4(x + 3y + 4) = 0$

14) $y - 6 = 2x + 5$

15) $4(y + 2) = 3(x - 2)$

16) $\frac{2}{5}y + \frac{3}{5}x + \frac{4}{5} = 0$

Point–slope Form

Find the slope of the following lines. Name a point on each line.

1) $y = 3(x + 5)$

2) $y + 2 = \dfrac{1}{4}(x - 3)$

3) $y + 1 = -2.5x$

4) $y - 4 = \dfrac{1}{3}(x - 4)$

5) $y + 5 = 0.6(x + 7)$

6) $y - 6 = -2x$

7) $y - 10 = -2(x - 9)$

8) $y + 18 = 0$

9) $y + 19 = 6(x + 1)$

10) $y - 14 = -3(x - 2)$

Write an equation in point–slope form for the line that passes through the given point with the slope provided.

11) $(9, -7), m = 5$

12) $(-2, 5), m = \dfrac{1}{2}$

13) $(0, -4), m = -3$

14) $(-a, b), m = n$

15) $(-8, 2), m = 4$

16) $(6, 1), m = -4$

17) $(-7, 12), m = \dfrac{1}{6}$

18) $(0, 13), m = 0$

19) $\left(-\dfrac{1}{2}, 2\right), m = \dfrac{1}{7}$

20) $(0, 0), m = -2$

Graphing Lines of Equations

✏️ Sketch the graph of each line

1) $y = 3x - 2$

2) $y = -\frac{1}{2}x + \frac{3}{2}$

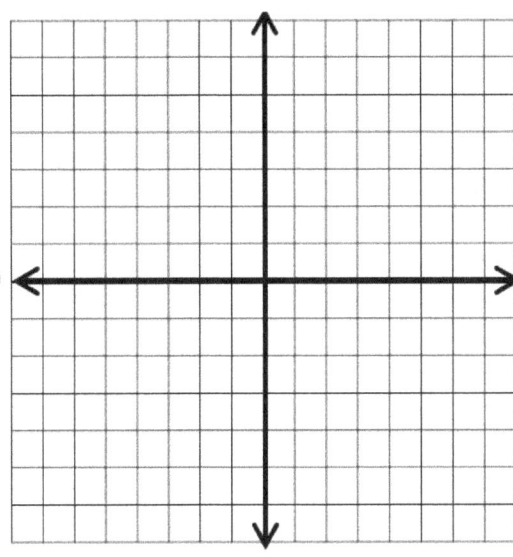
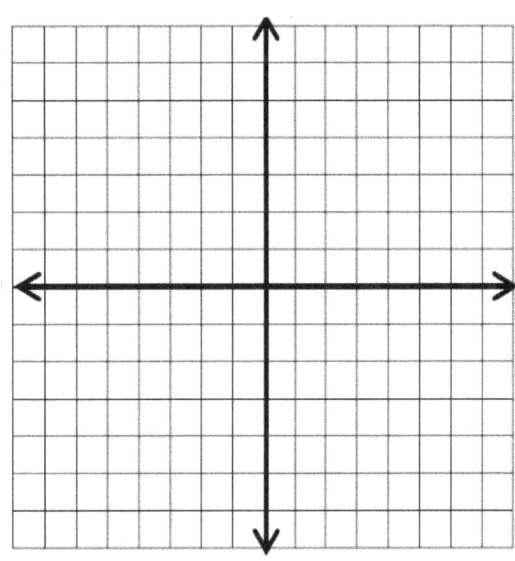

3) $4x - 5y = 12$

4) $-3x - y = 5$

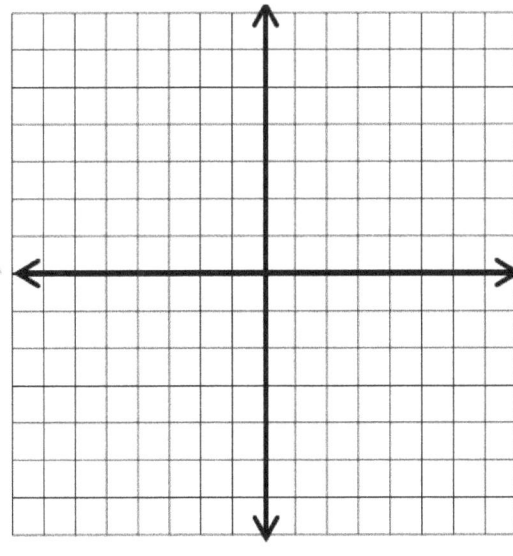
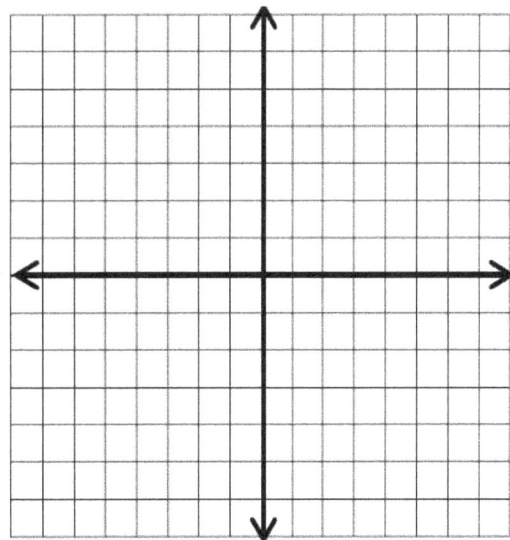

Equation of Parallel or Perpendicular Lines

🖉 **Write an equation of the line that passes through the given point and is parallel to the given line.**

1) $(-3, -1), x + 2y = -10$

2) $(-3, 2), y = x - 4$

3) $(-3, 1), 4y = x - 7$

4) $(0, 1), -y + 2x - 8 = 0$

5) $(2, 8), y + 12 = 0$

6) $(1, 4), -4x - 2y = -5$

7) $(-3, 0), y = \frac{2}{3}x + 4$

8) $(-1, 3), -4x + y = -16$

9) $(1, -1), y = -\frac{1}{5}x - 2$

10) $(-3, -3), 2x + 10y = -20$

🖉 **Write an equation of the line that passes through the given point and is perpendicular to the given line.**

11) $(-4, 0), 2x + y = -8$

12) $(-\frac{1}{2}, \frac{3}{4}), 9x - 6y = -9$

13) $(4, -8), y = -8$

14) $(9, -5), x = 9$

15) $(-8, 7), y = \frac{1}{4}x + 9$

16) $(\frac{1}{3}, \frac{2}{3}), y = -4x + 2$

17) $(-8, -4), y = \frac{7}{4}x + 10$

18) $(-8, 5), y = x + 13$

19) $(-4, -10), y = \frac{9}{4}x - 1$

20) $(5, 2), 5y - x + 8 = 13$

Equations of Horizontal and Vertical Lines

✎ Sketch the graph of each line.

1) $y = -2$

2) $y = 1$

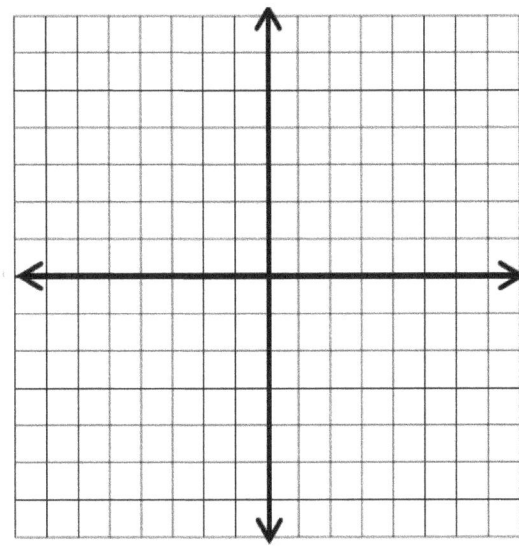
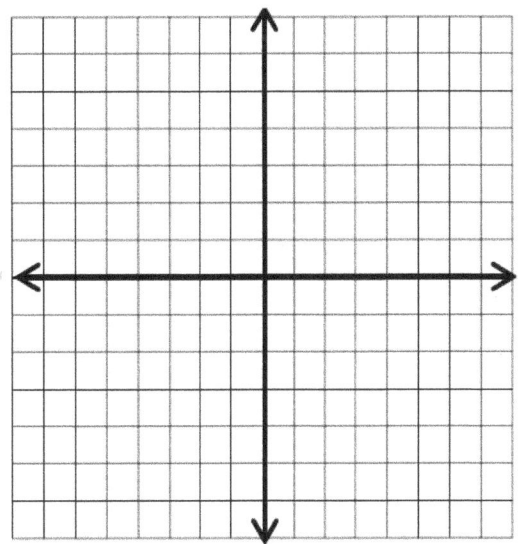

3) $x = -1$

4) $x = 2$

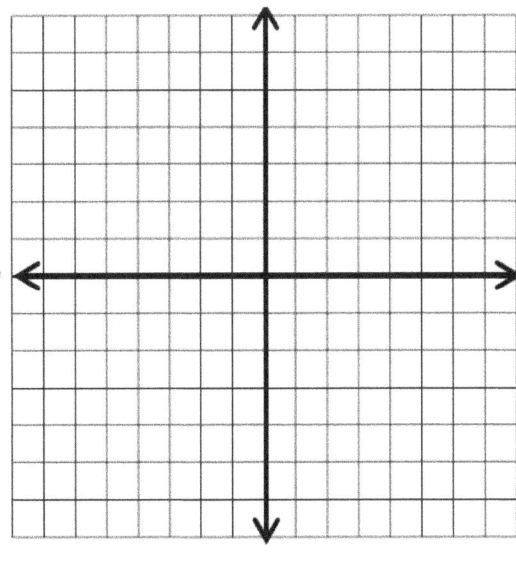
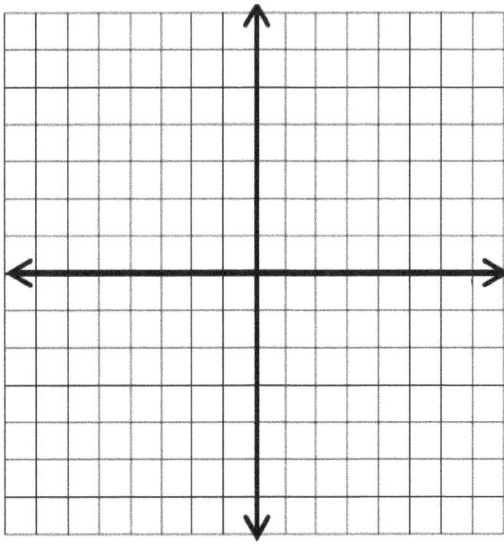

Graphing Absolute Value Equations

🖎 **Graph each equation.**

1) $y = |x + 4|$ 2) $y = |x + 1|$ 3) $y = -|x| - 1$

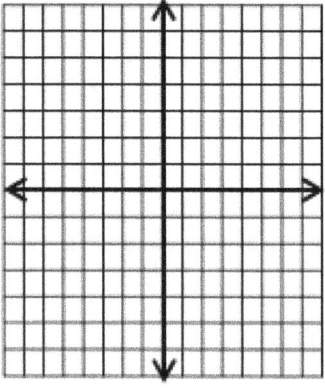 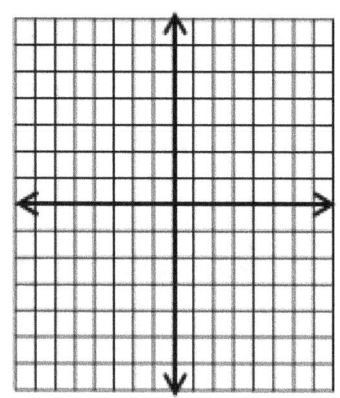 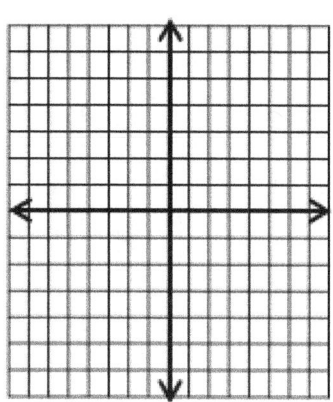

4) $y = |x - 2|$ 5) $y = -|x - 2|$ 6) $y = -2|2x + 2| + 4$

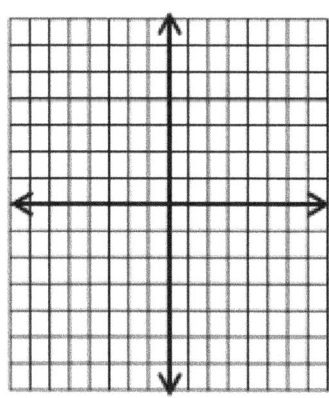 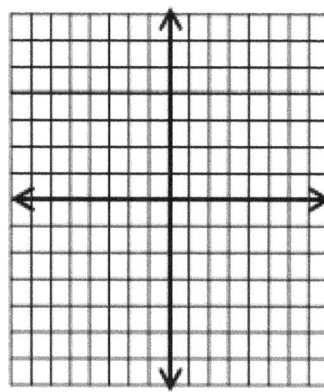 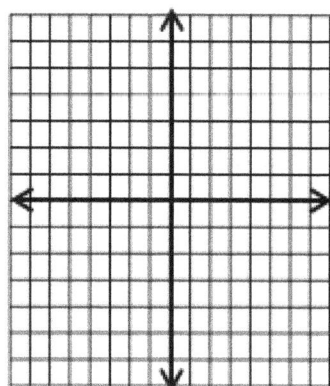

Answers of Worksheets – Chapter 8

Relation and Functions

1) No, $D_f = \{2, 4, 8, 9, 10\}$, $R_f = \{1, 0, 5, 7, 14\}$
2) Yes, $D_f = \{2, 1, -1, 7, 9\}$, $R_f = \{3, 0, -4, 5\}$
3) Yes, $D_f = (-\infty, \infty)$, $R_f = \{2, -\infty)$
4) No, $D_f = \{6, 3, 2, 5\}$, $R_f = \{-8, -2, 4, 0, 9\}$
5) No, $D_f = [-2, \infty)$, $R_f = (-\infty, \infty)$
6) No, $D_f = \{3, 5, 6\}$, $R_f = \{0, 7, 1, 9\}$

Finding Slope

1) 2
2) -1
3) 4
4) -3
5) 7
6) -4
7) 7
8) -8
9) -9
10) 8
11) $\frac{1}{5}$
12) $-\frac{3}{7}$
13) $\frac{1}{2}$
14) -1
15) $\frac{3}{8}$
16) $\frac{3}{11}$
17) 2
18) 1
19) $-\frac{5}{2}$
20) 3
21) -1
22) 3
23) 2
24) -3
25) -1
26) 3
27) 3
28) -11

Graphing Lines Using Line Equation

1) $y = x - 5$ 2) $y = -3x + 4$ 3) $x - 2y = 0$

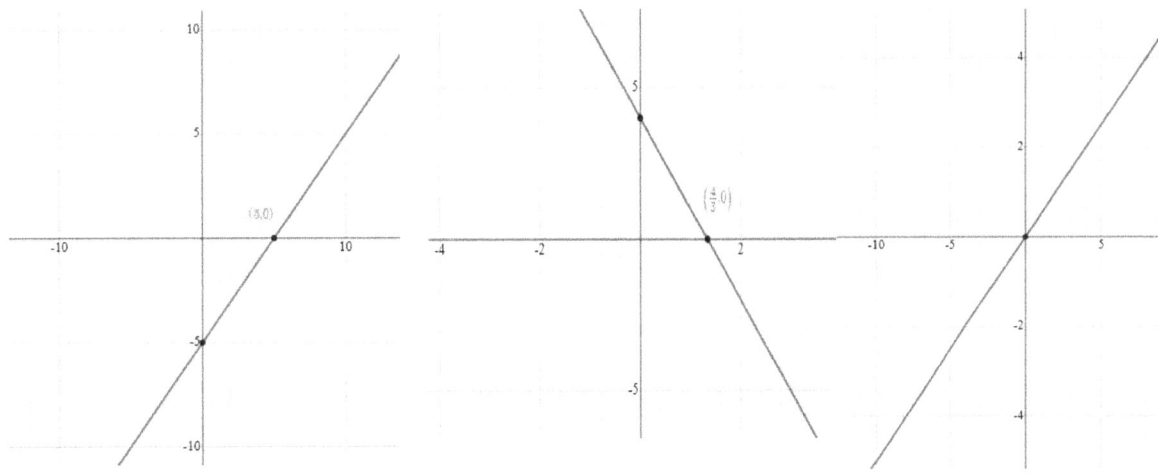

4) $x + y = -4$

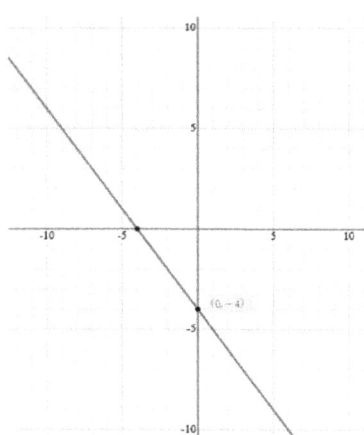

5) $4x + 3y = -2$

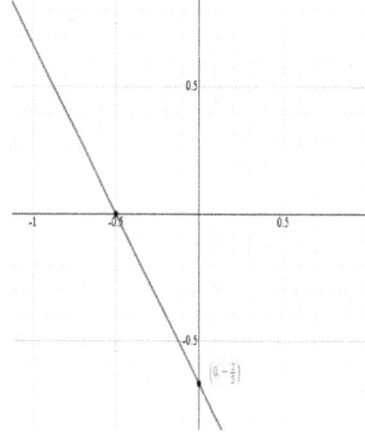

6) $y - 3x + 6 = 0$

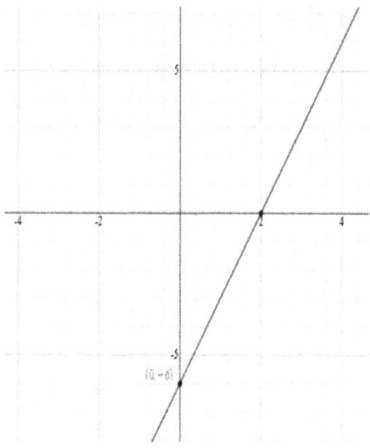

Writing Linear Equations

1) $y = 6x - 46$
2) $y = -3x + 34$
3) $y = x + 10$
4) $y = x - 4$
5) $y = -2x + 23$
6) $y = x - 15$
7) $y = -x + 20$
8) $y = -2x + 21$

9) $y = -3x + 42$
10) $y = -6x + 191$
11) $y = 8x + 57$
12) $y = 4x - 24$
13) $y = 4x - 34$
14) $y = -2x - 4$
15) $y = \frac{7}{6}x - \frac{5}{3}$

16) $y = -\frac{11}{3}x + \frac{28}{3}$
17) $y = 3x + 11$
18) $y = 5x + 15$
19) $y = 7x - 19$
20) $y = -3x - 1$
21) $y = -6x - 11$
22) $y = 5x + 22$

Graphing Linear Inequalities

1) $y > 3x - 5$

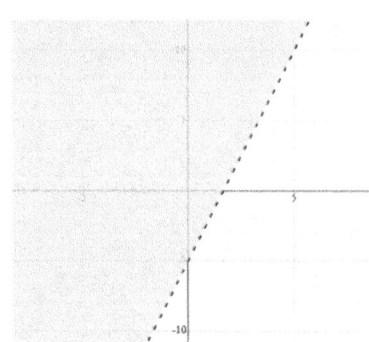

2) $y < 2x + 1$

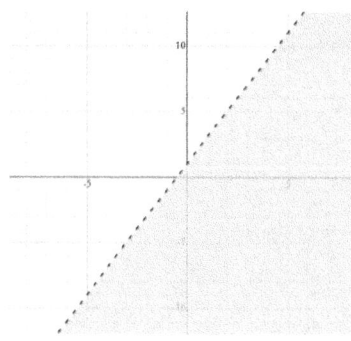

3) $y \leq -4x - 5$

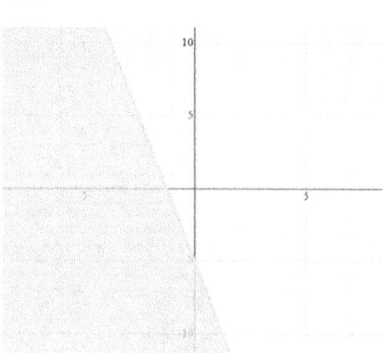

4) $2y \geq 12 + 4x$ 5) $-5y < x - 15$ 6) $3y \geq -9x + 6$

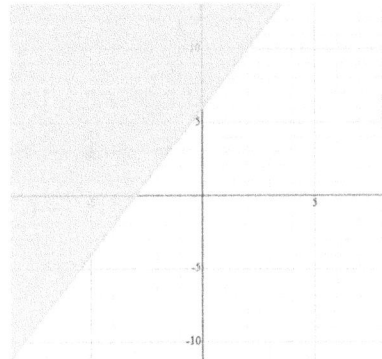
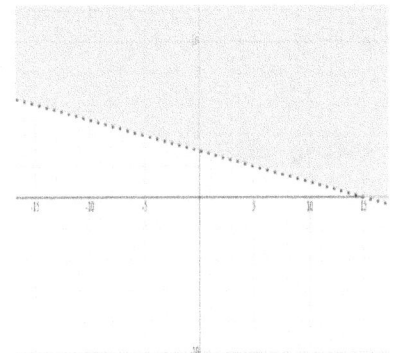
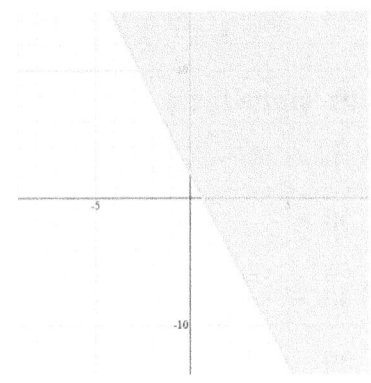

Write an equation from a graph

1) $y = 4x + 2$ 2) $y = 2x + 6$ 3) $y = 8x$ 4) $y = 3x + 1$

Rate of change

1) 14   2) -10 3) 2

x–intercept and y–intercept

1) $y - intercept = 6$ $x - intercept = 3$
2) $y - intercept = 4$ $x - intercept = -4$
3) $y - intercept = -18$ $x - intercept = 6$
4) $y - intercept = -6$ $x - intercept = -6$
5) $y - intercept = -\frac{8}{5}$ $x - intercept = \frac{8}{7}$
6) $y - intercept = -\frac{12}{5}$ $x - intercept = 3$
7) $y - intercept = -\frac{15}{4}$ $x - intercept = -\frac{5}{4}$
8) $y - intercept =$ undefind $x - intercept = 4$
9) $y - intercept = 4$ $x - intercept =$ undefind
10) $y - intercept = 6$ $x - intercept = 10$

Find the value of b

11) -11 12) 1   13) 3   14) 2

Slope–intercept form

1) $y = 15x + 7$ 4) $y = -\frac{1}{2}x - \frac{3}{2}$ 6) $y = \frac{7}{5}x + 2$ 8) $y = \frac{6}{5}x + \frac{36}{5}$

2) $y = -5x - 12$ 5) $y = \frac{2x}{5} - 3$ 7) $y = 3x + 6$ 9) $y = 12x - 20$

3) $y = -\frac{1}{3}x + 2$

10) $y = -\frac{5}{2}x + 4$

11) $y = \frac{3}{2}x + 4$

12) $y = -\frac{12}{7}x - 3$

13) $y = \frac{1}{3}x - \frac{4}{3}$

14) $y = 2x + 11$

15) $y = \frac{3}{4}x - \frac{7}{2}$

16) $y = -\frac{3}{2}x - 2$

Point–slope form

1) $m = 3, (-5, 0)$

2) $m = \frac{1}{4}, (3, -2)$

3) $m = -\frac{5}{2}, (0, -1)$

4) $m = 3, (4, 4)$

5) $m = \frac{6}{10}, (-7, -5)$

6) $m = -2, (0, 6)$

7) $m = -2, (9, 10)$

8) $m = 0, (0, -18)$

9) $m = 6, (-1, -19)$

10) $m = -3, (-2, 14)$

11) $y + 7 = 5(x - 9)$

12) $y - 5 = \frac{1}{2}(x + 2)$

13) $y + 4 = -3x$

14) $y - b = n(x + a)$

15) $y - 2 = 4(x + 8)$

16) $y - 1 = -4(x - 6)$

17) $y - 12 = \frac{1}{6}(x + 7)$

18) $y - 13 = 0$

19) $y - 2 = \frac{1}{7}\left(x + \frac{1}{2}\right)$

20) $y = -5x$

Graphing Line of Equation

1) $y = 3x - 2$

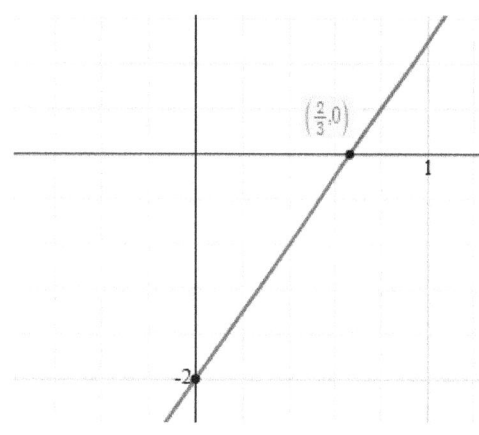

2) $y = -\frac{1}{2}x + \frac{3}{2}$

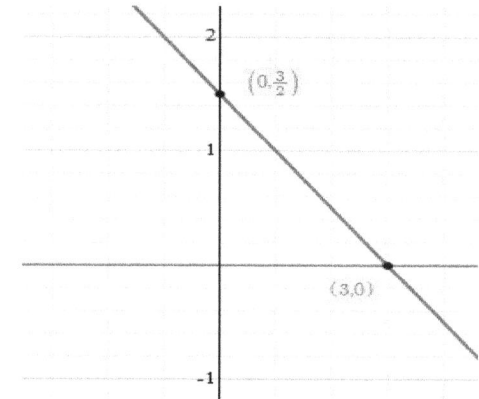

3) $4x - 5y = 12$

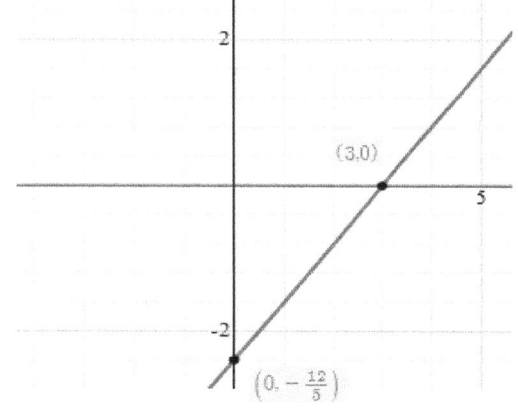

4) $-3x - y = 5$

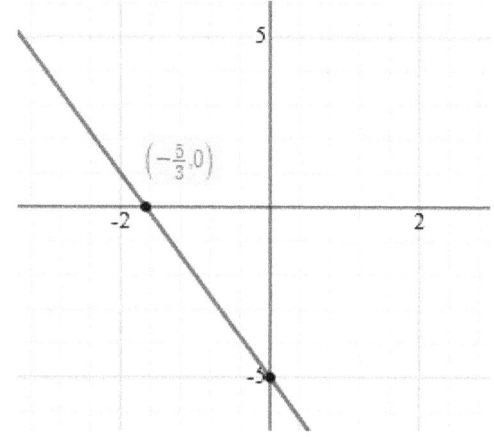

CLEP College Math Workbook

Equation of parallel or perpendicular line.

1) $y = -\frac{1}{2}x - 2\frac{1}{2}$
2) $y = x + 5$
3) $y = \frac{1}{4}x + \frac{7}{4}$
4) $y = 2x + 1$
5) $y = 8$
6) $y = -2x + 6$
7) $y = \frac{2}{3}x + 2$
8) $y = 4x + 7$
9) $y = -\frac{1}{5}x - \frac{4}{5}$
10) $y = -\frac{1}{5}x - \frac{18}{5}$
11) $y = \frac{1}{2}x + 2$
12) $y = -\frac{2}{3}x + \frac{5}{12}$
13) $x = 4$
14) $y = -5$
15) $y = -4x - 25$
16) $y = \frac{1}{4}x + \frac{7}{12}$
17) $y = -\frac{4}{7}x - \frac{60}{7}$
18) $y = -x - 3$
19) $y = -\frac{4}{9}x - \frac{106}{9}$
20) $y = -5x + 27$

Equations of horizontal and vertical lines

1) $y = -2$

2) $y = 1$

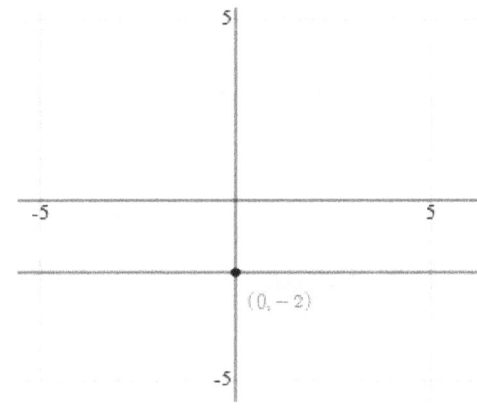
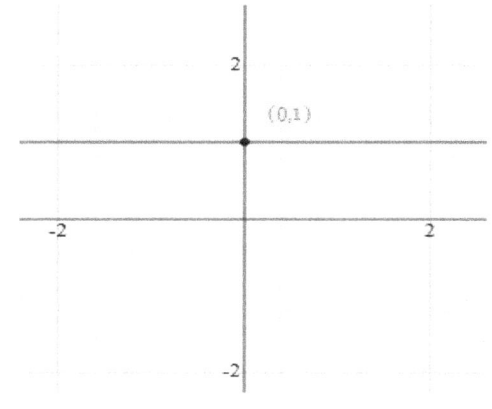

3) $x = -1$

4) $x = 2$

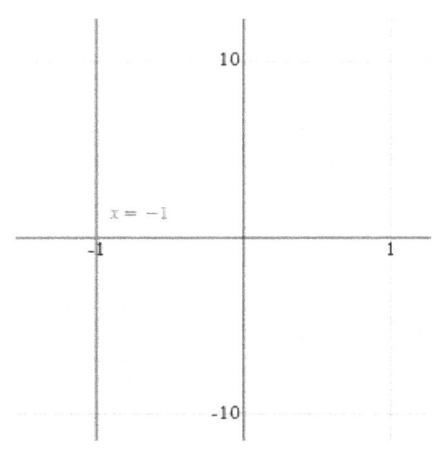
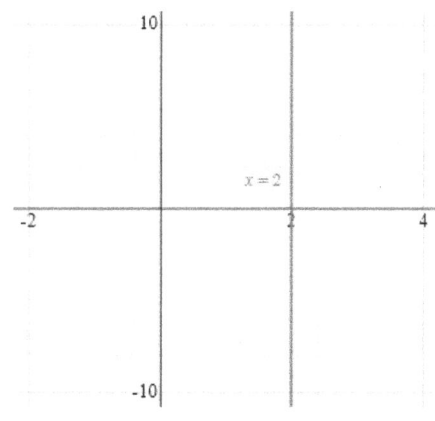

Graphing Absolute Value Equations

1) $y = |x + 4|$

2) $y = |x - 1|$

3) $y = -|x| - 1$

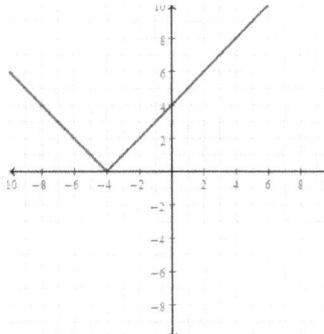
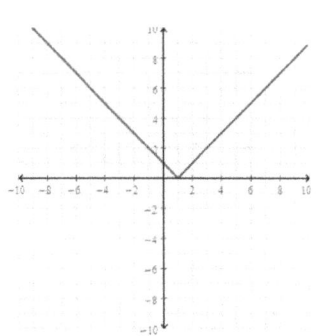
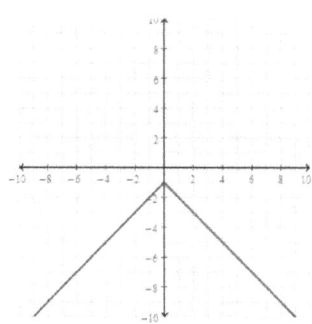

4) $y = |x - 2|$

5) $y = -|x - 2|$

6) $y = -2|2x + 2| + 4$

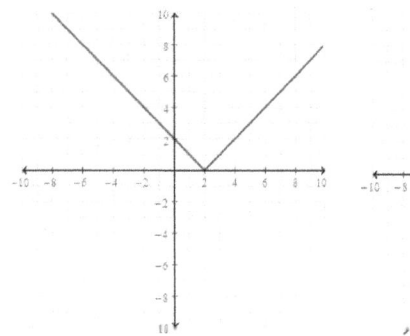
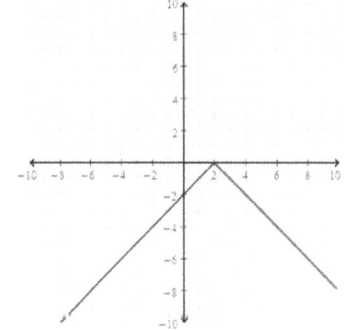
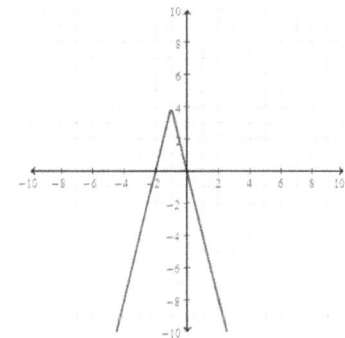

Chapter 9:
Monomials and polynomials

Topics that you will practice in this chapter:

- ✓ GCF of Monomials
- ✓ Factoring Quadratics
- ✓ Factoring by Grouping
- ✓ GCF and Powers of Monomials
- ✓ Writing Polynomials in Standard Form
- ✓ Simplifying Polynomials
- ✓ Adding and Subtracting Polynomials
- ✓ Multiplying a Polynomial and a Monomial
- ✓ Multiplying Binomials
- ✓ Factoring Trinomials
- ✓ Operations with Polynomials

Mathematics is, as it were, a sensuous logic, and relates to philosophy as do the arts, music, and plastic art to poetry. — K. Shegel

GCF of Monomials

✏️ **Find the GCF of each set of monomials.**

1) $39x, 30xy$

2) $60a, 56a^2$

3) $18x^2, 54x^2$

4) $36x^2, 21x^3$

5) $20a^2, 30a^2b$

6) $80a^3, 30a^2b$

7) $54x^3, 36x^3$

8) $33x, 44y^2x$

9) $15x^2, 12, 48$

10) $10v^3, 45v^3, 35v$

11) p^2q^2, pqr

12) $15m^2n, 25m^2n^2$

13) $12x^2yz, 3xy^2$

14) $22m^5n^2, 11m^2n^4$

15) $16x^3y, 8x^2$

16) $14ab^5, 7a^2b^2c$

17) $12t^7u^2, 18t^3u^7$

18) $18t, 48t^4$

19) $18r^3t, 26qr^2t^4$

20) $11a^4b^3, 44a^2b^5$

21) $16f, 21ab^2$

22) $12a^2b^2c^2, 20abc$

23) $18ab, 9ab$

24) $22m^5n^2, 11m^2n^4$

25) $4xy, 2x^2$

26) $x^3yz^2, 2x^3yz^3$

27) $140x, 140y^2, 80y^2$

28) $24a, 36a, 24ab^2$

29) $10x^3, 45x^3, 35x$

30) $105a, 30ab, 75a$

Factoring Quadratics

✎ **Factor each completely.**

1) $x^2 - 16x + 63 =$

2) $m^2 - 9m + 8 =$

3) $p^2 - 5p - 14 =$

4) $2b^2 + 17b + 21 =$

5) $a^2 + 5a + 4 =$

6) $a^2 + 2a - 15 =$

7) $4n^2 + 12n + 9 =$

8) $t^2 + 2t - 19 =$

9) $3x^3 + 21x^2 + 36x =$

10) $x^2 + 5x + 6 =$

11) $9r^2 - 5r - 10 =$

12) $30n^2b - 87nb + 30b =$

13) $7x^2 - 32x - 60 =$

14) $3b^3 - 5b^2 + 2b =$

15) $10m^2 + 89m - 9 =$

16) $4x^3 + 43x^2 + 30x =$

17) $9x^2 + 7 - 56 =$

18) $p^2 - 5p - 14 =$

19) $x^2 - 7x - 18 =$

20) $7x^2 - 31x - 20 =$

21) $6n^2 + 7n - 49 =$

22) $-6x^2 - 25x - 25 =$

23) $6x^2 + 5x - 6 =$

24) $16x^2 + 60x - 100 =$

25) $4x^2 - 35x + 49 =$

26) $5x^2 - 18x + 9 =$

27) $9n^2 + 66n + 21 =$

28) $3x^2 - 8x + 4 =$

29) $6x^2 - 36xy =$

30) $-6x^3 - 23x^2y - 10y^2x =$

31) $9a^2 + 9ab - 4b^2 =$

32) $4x^2 + 4xy - 35y^2 =$

33) $7x^2y - 27xy^2 + 18y^3 =$

34) $-2x^2 + 8xy + 64y^2 =$

35) $25mp^2 - 45mp =$

36) $14b^2 + 142b + 144 =$

37) $5x^2 + 85xy + 350y^2 =$

38) $7x^2 + 9xy =$

Factoring by Grouping

✎ Factor each completely.

1) $28xy - 7k - 49x + 4ky =$

2) $7xy - 3n - x + 21ny =$

3) $56n^3 + 64n^2 + 70n + 80 =$

4) $32u^2v - 12u^3m + 48u^4 - 8umv =$

5) $70n^4 + 40n^3 + 28n^2 + 16n =$

6) $45uv - 125bu - 75u^2 + 75bv =$

7) $x^3 + 7x^2 + 6x + 42 =$

8) $6x^3 + 36x^2 + 30x + 180 =$

9) $6m^3 - 30m^2 + 30m - 150 =$

10) $2x^3 - 4x^2 - 10x + 20 =$

11) $24p^3 + 15p^2 - 56p - 35 =$

12) $42mc + 36md - 7n^2c - 6n^2d =$

13) $28x^4 + 112x^2 - 21x^2 - 84x =$

14) $15xw + 18xk + 25yw + 30k =$

15) $56xy - 35x + 16ry - 10r =$

16) $4xy + 6 - x - 24y =$

17) $192x3 + 72x2 + 144x + 54 =$

18) $8x3 - 8x2 + 14x - 14 =$

19) $20x^3 + 5x^2 + 28x + 7 =$

20) $100x^3 + 160x^2 - 60x - 96 =$

GCF and Powers of Monomials

Find the GCF of each pairs of expressions.

1) $54x^3, 36x^3$
2) 2) $33x, 44y^2x$
3) $15x^2, 12, 48$
4) 4) $10v^3, 45v^3, 35v$
5) p^2q^2, pqr
6) 6) $15m^2n, 25m^2n^2$
7) $12x^2yz, 3xy^2$
8) 8) $22m^5n^2, 11m^2n^4$
9) $16x^3y, 8x^2$
10) $14ab^5, 7a^2b^2c$

11) $12t^7u^2, 18t^3u^7$
12) 12) $18t, 48t^4$
13) $18r^3t, 26qr^2t^4$
14) 14) $11a^4b^3, 44a^2b^5$
15) $16f, 21ab^2$
16) 16) $12a^2b^2c^2, 20abc$
17) $18ab, 9ab$
18) 18) $22m^5n^2, 11m^2n^4$
19) $4xy, 2x^2$
20) 20) $x^3yz^2, 2x^3yz^3$

Simplify.

21) $(3x^4)^7$
22) $(4y^22y^3y)^2$
23) $(3x^2\,2x^2)^3$
24) $(8x^4y^3)^6$
25) $(3y^25y^2)^2$
26) $(6x^3y)^3$
27) $(8x^2x^23n)^2$
28) $(7xy^6)^3$
29) $(9x^3y^2)^4$

30) $(10y^3y^2)^3$
31) $(6x^2x^6)^3$
32) $(3x^74x^3k^2)^2$
33) $(4y^54y^2)^2$
34) $(5x2x^3)^3$
35) $(4y^3)^3$
36) $(y^3y^3y^2)^3$
37) $(4y^2y)^3$
38) $(6xy^6)^3$

Writing Polynomials in Standard Form

✎ Write each polynomial in standard form.

1) $9x - 7x =$

2) $-6 + 15x - 15x =$

3) $3x^2 - 11x^3 =$

4) $18 + 19x^3 - 14 =$

5) $3x^2 + 9x - 4x^5 =$

6) $-7x^3 + 12x^7 =$

7) $9x + 6x^2 - 2x^6 =$

8) $-5x^3 + x - 9x^4 =$

9) $8x^2 + 34 - 21x =$

10) $8 - 7x + 11x^4 =$

11) $25x^3 + 45x - 13x^4 =$

12) $17 + 9x^2 - 2x^3 =$

13) $18x^2 - 8x + 8x^3 =$

14) $9x^4 - 4x^2 - 10x^5 =$

15) $-41 + 7x^2 - 8x^4 =$

16) $8x^2 - 7x^5 + 3x^3 - 12 =$

17) $4x^2 - 9x^5 + 12 - 8x^4 =$

18) $-2x^5 + 6x - 9x^2 - 7x =$

19) $14x^5 + 7x^4 - 8x^5 - 8x^2 =$

20) $2x^3 - 15x^4 + 9x^3 + 3x^8 =$

21) $7x^4 - 16x^5 - 9x^2 + 10x^4 =$

22) $5x^2 + 6x^5 + 37x^3 - 9x^5 =$

23) $3x(2x + 5 - 6x^2) =$

24) $12x(x^6 + 2x^3) =$

25) $6x(x^2 + 8x + 4) =$

26) $8x(3 - 2x + 4x^3) =$

27) $7x(2x^3 - 2x^2 + 2) =$

28) $5x(5x^5 + 4x^4 - 1) =$

29) $x(4x^3 + 52x^4 + 2x) =$

30) $6x(3x - 4x^4 + 7x^2) =$

Simplifying Polynomials

✎ **Simplify each expression.**

1) $3(x - 12) =$

2) $5x(2x - 4) =$

3) $7x(5x - 1) =$

4) $6x(3x + 2) =$

5) $5x(2x - 7) =$

6) $9x(x + 8) =$

7) $(3x - 8)(x - 3) =$

8) $(x - 9)(3x + 4) =$

9) $(x - 8)(x - 5) =$

10) $(3x + 4)(3x - 4) =$

11) $(5x - 8)(5x - 2) =$

12) $7x^2 + 7x^2 - 6x^4 =$

13) $5x - 2x^2 + 7x^3 + 10 =$

14) $8x + 2x^2 - 5x^3 =$

15) $15x + 4x^5 - 8x^2 =$

16) $-4x^2 + 7x^5 + 11x^4 =$

17) $-14x^2 + 8x^3 - 2x^4 + 5x =$

18) $14 - 5x^2 + 6x^2 - 10x^3 + 17 =$

19) $x^2 - 9x + 2x^3 + 15x - 10x =$

20) $14 - 8x^2 + 4x^2 - 9x^3 + 1 =$

21) $-4x^5 + 2x^4 - 18x^2 + 2x^5 =$

22) $(3x^3 - 5) + (3x^3 - 2x^3) =$

23) $4(3x^5 - 3x^3 - 6x^5) =$

24) $-4(x^5 + 8) - 4(12 - x^5) =$

25) $7x^2 - 9x^3 - 2x + 14 - 5x^2 =$

26) $10 - 5x^2 + 3x^2 - 4x^3 + 4 =$

27) $(8x^2 - 2x) - (5x - 5 - 4x^2) =$

28) $4x^4 - 8x^3 - x(3x^2 + 5x) =$

29) $4x + 8x^2 - 10 - 2(x^2 - 1) =$

30) $5 - 3x^2 + (6x^4 - 2x^2 + 8x^4) =$

31) $-(x^5 + 8) - 7(4 + x^5) =$

32) $(4x^3 - x) - (x - 6x^3) =$

Adding and Subtracting Polynomials

✎ **Add or subtract expressions.**

1) $(-x^3 - 3) + (4x^3 + 2) =$

2) $(3x^2 + 4) - (6 - x^2) =$

3) $(x^3 + 4x^2) - (5x^3 + 15) =$

4) $(3x^3 - 2x^2) + (2x^2 - x) =$

5) $(10x^3 + 14x) - (14x^3 + 7) =$

6) $(5x^2 - 7) + (3x^2 + 7) =$

7) $(9x^3 + 4) - (10 - 5x^3) =$

8) $(x^2 + 2x^3) - (2x^3 + 5) =$

9) $(8x^2 - x) + (5x - 4x^2) =$

10) $(17x + 10) - (2x + 10) =$

11) $(12x^4 - 4x) - (x - 3x^4) =$

12) $(3x - x^4) - (7x^4 + 8x) =$

13) $(7x^3 - 6x^5) - (4x^5 - 2x) =$

14) $(x^3 - 7) + (4x^3 + 8x^5) =$

15) $(6x^2 + 5x^4) - (x^4 - 9x^2) =$

16) $(-4x^2 - 4x) + (7x - 8x^2) =$

17) $(x - 6x^4) - (15x^4 + 2x) =$

18) $(4x - 3x^4) - (2x^4 - 3x^3) =$

19) $(7x^3 - 7) + (6x^3 - 6x^2) =$

20) $(9x^5 + 7x^4) - (x^4 - 5x^5) =$

21) $(-4x^2 + 11x^4 + 2x^3) + (20x^3 + 4x^4 + 12x^2) =$

22) $(5x^2 - 5x^4 - 5x) - (-4x^2 - 5x^4 + 5x) =$

23) $(12x + 36x^3 - 10x^4) + (20x^3 + 10x^4 - 7x) =$

24) $(2x^5 - 4x^3 - 5x) - (2x^2 + 7x^3 - 2x) =$

25) $(14x^3 - 4x^5 - x) - (-4x^3 - 12x^5 + 9x) =$

26) $(-5x^2 + 12x^4 + x^3) + (10x^3 + 17x^4 + 7x^2) =$

Multiplying a Polynomial and a Monomial

✍ **Find each product.**

1) $2x(x+4) =$

2) $3(8-x) =$

3) $5x(3x+4) =$

4) $x(-2x+5) =$

5) $7x(3x-3) =$

6) $3(2x-5y) =$

7) $6x(7x-3) =$

8) $x(12x+5y) =$

9) $5x(x+6y) =$

10) $11x(4x+5y) =$

11) $8x(4x+2) =$

12) $12x(x-15y) =$

13) $9x(5x-3y) =$

14) $8x(5x-2y+5) =$

15) $9x(2x^2+7y^2) =$

16) $8x(9x+6y) =$

17) $2(3x^5-2y^5) =$

18) $4x(-x^2y+2y) =$

19) $-3(2x^3-3xy+9) =$

20) $2(x^2-2xy-4) =$

21) $7x(4x^3-xy+2x) =$

22) $-9x(-2x^3-2x+7xy) =$

23) $6(x^2+3xy-8y^2) =$

24) $5x(7x^3-x+8) =$

25) $7(x^{24}-4x-6) =$

26) $x^2(-3x^3+4x+7) =$

27) $x^2(2x^3+10-5x) =$

28) $4x^4(3x^3-2x+8) =$

29) $5x^2(x^4-5xy+2y^3) =$

30) $4x^2(7x^4-2x+11) =$

31) $7x^3(3x^3+5x-7) =$

32) $4x(x^2-8xy+7y^3) =$

Multiplying Binomials

✍ **Find each product.**

1) $(x+5)(x+1) =$

2) $(x-3)(x+7) =$

3) $(x-1)(x-9) =$

4) $(x+3)(x+8) =$

5) $(x-4)(x-11) =$

6) $(x+5)(x+6) =$

7) $(x-8)(x+7) =$

8) $(x-3)(x-2) =$

9) $(x+8)(x+11) =$

10) $(x-3)(x+5) =$

11) $(x+8)(x+8) =$

12) $(x+2)(x+7) =$

13) $(x-9)(x+4) =$

14) $(x-10)(x+10) =$

15) $(x+24)(x+2) =$

16) $(x+9)(x+13) =$

17) $(x-7)(x+7) =$

18) $(x-5)(x+2) =$

19) $(3x+4)(x+5) =$

20) $(x-8)(5x+2) =$

21) $(x-9)(4x+9) =$

22) $(2x-7)(3x-2) =$

23) $(x-4)(x+11) =$

24) $(5x-6)(2x+4) =$

25) $(4x-9)(x+7) =$

26) $(8x-5)(2x+2) =$

27) $(3x+9)(7x+4) =$

28) $(6x-8)(4x+4) =$

29) $(4x+5)(5x-8) =$

30) $(8x-1)(8x+4) =$

31) $(9x+4)(3x-6) =$

32) $(4x^2+12)(4x^2-12) =$

Factoring Trinomials

✎ **Factor each trinomial.**

1) $x^2 + 12x + 35 =$

2) $x^2 - 8x + 12 =$

3) $x^2 + 11x + 10 =$

4) $x^2 - 12x + 27 =$

5) $x^2 - 16x + 15 =$

6) $x^2 - 13x + 40 =$

7) $x^2 + 15x + 44 =$

8) $x^2 + x - 72 =$

9) $x^2 - 81 =$

10) $x^2 - 17x + 70 =$

11) $x^2 + 8x - 48 =$

12) $x^2 + 5x - 104 =$

13) $x^2 - 7x - 18 =$

14) $x^2 + 22x + 121 =$

15) $3x^2 - 3x - 36 =$

16) $2x^2 - 35x + 75 =$

17) $14x^2 + 11x - 15 =$

18) $8x^2 - 12x - 20 =$

19) $15x^2 + 16x + 4 =$

20) $24x^2 + 2x - 1 =$

✎ **Calculate each problem.**

21) The area of a rectangle is $x^2 - 3x - 40$. If the width of rectangle is $x - 8$, what is its length? _____

22) The area of a parallelogram is $12x^2 + 7x - 10$ and its height is $4x + 5$. What is the base of the parallelogram? _____

23) The area of a rectangle is $10x^2 - 43x + 28$. If the width of the rectangle is $5x - 4$, what is its length? _____

Operations with Polynomials

✎ **Find each product.**

1) $2(4x + 1) =$ _____

2) $5(2x + 7) =$ _____

3) $4(6x - 5) =$ _____

4) $-4(7x - 8) =$ _____

5) $3x^2(8x + 4) =$ _____

6) $6x^2(2x - 9) =$ _____

7) $5x^3(-x + 4) =$ _____

8) $-5x^4(4x - 9) =$ _____

9) $6(x^2 + 7x - 3) =$ _____

10) $4(3x^2 - 2x + 6) =$ _____

11) $9(3x^2 + 8x + 2) =$ _____

12) $7x(x^2 + 5x + 3) =$ _____

13) $(7x + 2)(2x - 5) =$ _____

14) $(8x + 5)(3x - 8) =$ _____

15) $(4x + 2)(6x - 1) =$ _____

16) $(5x - 4)(5x + 9) =$ _____

✎ **Calculate each problem.**

17) The measures of two sides of a triangle are $(2x + 8y)$ and $(5x - 3y)$. If the perimeter of the triangle is $(11x + 6y)$, what is the measure of the third side? _____

18) The height of a triangle is $(8x + 2)$ and its base is $(2x - 6)$. What is the area of the triangle? _____

19) One side of a square is $(4x + 3)$. What is the area of the square? _____

20) The length of a rectangle is $(7x - 9y)$ and its width is $(13x + 9y)$. What is the perimeter of the rectangle? _____

21) The side of a cube measures $(x + 2)$. What is the volume of the cube? _____

22) If the perimeter of a rectangle is $(24x + 10y)$ and its width is $(4x + 3y)$, what is the length of the rectangle? _____

Answers of Worksheets – Chapter 9

GCF of Monomials

1) $3x$
2) 4
3) $18x^2$
4) $3x^2$
5) $10a^2$
6) $10a^2$
7) $18x^3$
8) $11x$
9) 3
10) $5v$
11) pq
12) $5m^2n$
13) $3xy$
14) $11m^2n$
15) $8x^2$
16) $7ab^2$
17) $6t^3u^2$
18) $8t$
19) $2r^2t$
20) $11a^2b^3$
21) no
22) $4abc$
23) $9ab$
24) $11m^2n^2$
25) $2x$
26) x^3yz^2
27) 20
28) $12a$
29) $5x$
30) $15a$

Factoring Quadratics

1) $(x-9)(x-7)$
2) $(m-1)(m-8)$
3) $(p+2)(p-7)$
4) $(2b+3)(b+7)$
5) $a^2 + 5a + 4$
6) $a^2 + 2a - 15$
7) $4n^2 + 12n + 9$
8) $t^2 + 2t - 19$
9) $3x^3 + 21x^2 + 36x$
10) $x^2 + 5x + 6$
11) $9r^2 - 5r - 10$
12) $30n^2b\ 87nb + 30b$
13) $7x^2 - 32x - 60$
14) $3b^3 - 5b^2 + 2b$
15) $10m^2 + 89m - 9$
16) $4x^3 + 43x^2 + 30x$
17) $9x^2 + 7x - 56$
18) $p^2 - 5p - 14$
19) $x^2 - 7x - 18$
20) $7x^2 - 31x - 20$
21) $(3n-7)(2x+7)$
22) $-(2x+5)(3x+5)$
23) $(2x+3)(3x-2)$
24) $4(x+5)(4x-5)$
25) $(x-7)(4x-7)$
26) $(5x-3)(x-3)$
27) $3(3n+1)(n+7)$
28) $(3x-2)(x-2)$
29) $6x(x-6y)$
30) $-x(2x+y)(3x+10y)$
31) $(3a+4b)(3a-b)$
32) $(2x+7y)(2x-5y)$
33) $y(7x-6y)(x-3y)$
34) $-2(x-8y)(x+4y)$
35) $5mp(5p-9)$
36) $2(7b+8)(b+9)$

CLEP College Math Workbook

37) $5(x + 10y)(x + 7y)$ 38) $x(7x + 9y)$

Factoring by Grouping

1) $(7x + k)(4y - 7)$
2) $(x + 3n)(7y - 1)$
3) $2(4n^2 + 5)(7n + 8)$
4) $4u(4u - m)(2v + 3u^2)$
5) $2n(5n^2 + 2)(7n + 4)$
6) $5(3u + 5b)(3v - 5u)$
7) $(x^2 + 6)(x + 7)$
8) $6(x^2 + 5)(x + 6)$
9) $6(m^2 + 5)(m - 5)$
10) $2(x^2 - 5)(x - 2)$
11) $(3p^2 - 7)(8p + 5)$
12) $(6m - n^2)(7c + 6d)$
13) $7x(4x^2 - 3)(x + 4)$
14) $(3x + 5y)(5w + 6k)$
15) $(7x + 2r)(8y - 5)$
16) $(x - 6)(4y - 1)$
17) $(4x^2 + 1)(3x - 5)$
18) $2(4x^2 + 7)(x - 1)$
19) $(5x^2 + 7)(4x + 1)$
20) $4(5x^2 - 3)(5x + 8)$

GCF and Powers of monomials

1) $10x^2$
2) 4
3) 10
4) $7xy$
5) $6y7x$
6) $3x$
7) 3
8) $7xy^2$
9) $18x^2$
10) $(15)x$
11) 27
12) 3
13) $20x$
14) $2x^2y$
15) xy
16) $2x^2y^2$
17) $6xy^4$
18) $5x^3y$
19) $2187x^{28}$
20) $72x^{12}$
21) $216x^{12}$
22) $262144x^{24}y^{18}$
23) $225y^{10}$
24) $216x^9y^3$
25) $576x^8n^2$
26) $343x^3y^{18}$
27) $6561x^{12}y^8$
28) $1000y^{15}$
29) $396x^{24}$
30) $144x^{100}k^4$
31) $256y^{14}$
32) $1000x^{12}$
33) $64y^9$
34) $27y^{18}$
35) $64y^9$
36) $216x^3y^{18}$

Writing Polynomials in Standard Form

1) $2x$
2) -6
3) $-11x^3 + 3x^2$
4) $19x^4 + 4$
5) $-4x^5 + 3x^2 + 9x$
6) $12x^7 - 7x^3$
7) $-2x^6 + 6x^2 + 9x$
8) $-9x^4 - 5x^3 + x$

9) $8x^2 - 21x + 34$

10) $11x^4 - 7x + 8$

11) $-13x^4 + 25x^3 + 45x$

12) $-2x^3 + 9x^2 + 17$

13) $8x^3 + 18x^2 - 8x$

14) $-10x^5 - 9x^4 - 4x^2$

15) $-8x^4 + 7x^2 - 41$

16) $-7x^5 + 3x^3 + 8x^2 - 12$

17) $-9x^5 - 8x^4 + 4x^2 + 12$

18) $-2x^5 - 9x^2 - x$

19) $6x^5 + 7x^4 - 8x^2$

20) $3x^8 - 15x^4 + 11x^2$

21) $-16x^5 + 17x^4 - 9x^2$

22) $-3x^5 + 37x^3 + 5x^2$

23) $-18x^3 + 6x^2 + 15x$

24) $12x^7 + 24x^4$

25) $6x^3 + 48x^2 + 24x$

26) $32x^4 - 16x^2 + 24x$

27) $14x^4 - 14x^3 + 14x$

28) $25x^6 + 20x^5 - 5x$

29) $52x^5 + 4x^4 + 2x^2$

30) $-24x^5 + 42x^3 + 18x^2$

Simplifying Polynomials

1) $3x - 36$

2) $10x^2 - 20x$

3) $35x^2 - 7x$

4) $18x^2 + 12x$

5) $10x^2 - 35x$

6) $9x^2 + 72x$

7) $3x^2 - 17x + 24$

8) $3x^2 - 23x - 36$

9) $x^2 - 13x + 40$

10) $9x^2 - 16$

11) $25x^2 - 50x + 16$

12) $-6x^4 + 14x^2$

13) $7x^3 - 2x^2 + 5x + 10$

14) $-5x^3 + 2x^2 + 8x$

15) $4x^5 - 8x^2 + 15x$

16) $7x^5 + 11x^4 - 4x^2$

17) $-2x^4 + 8x^3 - 14x^2 + 5x$

18) $-10x^3 + x^2 + 31$

19) $2x^3 + x^2 - 4x$

20) $-9x^3 - 4x^2 + 15$

21) $-2x^5 + 2x^4 - 18x^2$

22) $4x^3 - 5$

23) $-12x^5 - 12x^3$

24) -80

25) $-9x^3 + 2x^2 - 2x + 14$

26) $-4x^3 - 2x^2 + 14$

27) $12x^2 - 7x + 5$

28) $4x^4 - 11x^3 - 5x^2$

29) $6x^2 + 4x - 8$

30) $14x^4 - 5x^2 + 5$

31) $-8x^5 - 36$

32) $10x^3 - 2x$

Adding and Subtracting Polynomials

1) $3x^2 - 1$

2) $4x^2 - 2$

3) $-4x^3 + 4x^2 - 15$

4) $3x^3 - x$

5) $-4x^3 + 14x - 7$

6) $8x^2$

7) $14x^3 - 6$
8) $x^2 - 5$
9) $4x^2 + 4x$
10) $15x$
11) $15x^4 - 5x$
12) $-8x^4 - 5x$
13) $-10x^5 + 7x^3 + 2x$

14) $5x^5 + 5x^3 - 7$
15) $4x^4 + 15x^2$
16) $-12x^2 + 3x$
17) $-21x^4 - x$
18) $-5x^4 + 3x^3 + 4x$
19) $13x^3 - 6x^2 - 7$
20) $14x^5 + 6x^4$

21) $15x^4 + 22x^3 + 8x^2$
22) $9x^2 - 10x$
23) $56x^3 + 5x$
24) $2x^5 - 11x^3 - 2x^2 - 3x$
25) $8x^5 + 18x^3 - 10x$
26) $29x^4 + 11x^3 + 2x^2$

Multiplying a Polynomial and a Monomial

1) $2x^2 + 8x$
2) $-3x + 24$
3) $15x^2 + 20x$
4) $-2x^2 + 5x$
5) $21x^2 - 21x$
6) $6x - 15y$
7) $42x^2 - 18x$
8) $12x^2 + 5xy$
9) $5x^2 + 30xy$
10) $44x^2 + 55xy$
11) $32x^2 + 16x$
12) $12 - 180xy$
13) $45x^2 - 27xy$
14) $40x^2 - 16xy + 40x$
15) $18x^3 + 63xy^2$
16) $72x^2 + 48xy$

17) $6x^5 - 2y^5$
18) $-4x^3y + 8xy$
19) $-6x^3 + 9xy - 27$
20) $2x^2 - 4xy - 8$
21) $28x^4 - 7x^2y + 14x^2$
22) $18x^4 + 18x^2 - 63x^2y$
23) $6x^2 + 18xy - 48y^2$
24) $35x^4 - 5x^2 + 40x$
25) $7x^{24} - 28x - 42$
26) $-3x^5 + 4x^3 + 7x^2$
27) $2x^5 - 5x^3 + 10x^2$
28) $12x^7 - 8x^5 + 32x^4$
29) $5x^6 - 25x^3y + 10x^2y^3$
30) $28x^6 - 8x^3 + 44x^2$
31) $21x^6 + 35x^4 - 49x^3$
32) $4x^3 - 32x^2y + 28xy^3$

Multiplying Binomials

1) $x^2 + 6x + 5$
2) $x^2 + 4x - 21$
3) $x^2 - 10x + 9$
4) $x^2 + 11x + 24$
5) $x^2 - 15x + 44$
6) $x^2 + 11x + 30$

7) $x^2 - x - 56$
8) $x^2 - 5x + 6$
9) $x^2 + 19x + 88$
10) $x^2 + 2x + 15$
11) $x^2 + 16x + 64$
12) $x^2 + 9x + 14$

CLEP College Math Workbook

13) $x^2 - 5x - 36$
14) $x^2 - 100$
15) $x^2 + 26x + 48$
16) $x^2 + 22x + 117$
17) $x^2 - 49$
18) $x^2 - 3x - 10$
19) $3x^2 + 19x + 20$
20) $5x^2 - 38x - 16$
21) $4x^2 - 27x - 81$
22) $6x^2 - 25x + 14$

23) $x^2 + 7x - 44$
24) $10x^2 + 8x - 24$
25) $4x^2 + 19x - 63$
26) $16x^2 + 6x - 10$
27) $21x^2 + 75x + 36$
28) $24x^2 - 8x - 32$
29) $20x^2 - 7x - 40$
30) $64x^2 + 24x - 4$
31) $27x^2 - 42x - 24$
32) $16x^4 - 144$

Factoring Trinomials

1) $(x + 5)(x + 7)$
2) $(x - 2)(x - 6)$
3) $(x + 1)(x + 10)$
4) $(x - 9)(x - 3)$
5) $(x - 1)(x - 15)$
6) $(x - 5)(x - 8)$
7) $(x + 4)(x + 11)$
8) $(x + 9)(x - 8)$

9) $(x - 9)(x + 9)$
10) $(x - 7)(x - 10)$
11) $(x - 4)(x + 12)$
12) $(x - 8)(x + 13)$
13) $(x + 2)(x - 9)$
14) $(x + 11)(x + 11)$
15) $(3x + 9)(x - 4)$
16) $(x - 15)(2x - 5)$

17) $(7x - 5)(2x + 3)$
18) $(2x - 5)(4x + 4)$
19) $(3x + 2)(5x + 2)$
20) $(6x - 1)(4x + 1)$
21) $(x + 5)$
22) $(3x - 2)$
23) $(2x - 7)$

Operations with Polynomials

1) $8x + 2$
2) $10x + 35$
3) $24x - 20$
4) $-28x + 32$
5) $24x^3 + 12x^2$
6) $12x^3 - 54x^2$
7) $-5x^4 + 20x^3$
8) $-20x^5 + 45x^4$

9) $6x^2 + 42x - 18$
10) $12x^2 - 8x + 24$
11) $27x^2 + 72x + 18$
12) $7x^3 + 35x^2 + 21x$
13) $14x^2 - 31x - 10$
14) $24x^2 - 49x - 40$
15) $24x^2 + 8x - 2$
16) $25x^2 + 25x - 36$

17) $(4x + y)$
18) $8x^2 - 22x - 6$
19) $16x^2 + 24x + 9$
20) $40x$
21) $x^3 + 6x^2 + 12x + 8$
22) $(8x + 2y)$

Chapter 10: Functions Operations and Quadratic

Topics that you will practice in this chapter:

- ✓ Evaluating Functions
- ✓ Adding and Subtracting Functions
- ✓ Multiply and Dividing Functions
- ✓ Composition of Functions
- ✓ Solving Quadratic Equations
- ✓ Quadratic Formula and Discriminant
- ✓ Quadratic Inequalities
- ✓ Graphing Quadratic Functions
- ✓ Domain and Range of Radical Functions
- ✓ Solving Radical Equations

It's fine to work on any problem, so long as it generates interesting mathematics along the way – even if you don't solve it at the end of the day." – Andrew Wiles

Evaluating Function

✏️ **Write each of following in function notation.**

1) $h = -8x + 9$

2) $k = 5a - 21$

3) $d = 14t$

4) $y = \frac{3}{17}x - \frac{9}{17}$

5) $m = 18n - 94$

6) $c = p^2 - 7p + 15$

✏️ **Evaluate each function.**

7) $f(x) = 6x - 7$, find $f(-3)$

8) $g(x) = \frac{1}{10}x + 6$, find $f(5)$

9) $h(x) = -2x + 15$, find $f(8)$

10) $f(x) = -3x + 8$, find $f(-2)$

11) $f(a) = 12a - 9$, find $f(0)$

12) $h(x) = 18 - 5x$, find $f(-4)$

13) $g(n) = 7n - 5$, find $f(5)$

14) $f(x) = -9x - 2$, find $f(3)$

15) $k(n) = -12 + 4.5n$, find $f(2)$

16) $f(x) = -1.5x + 2.5$, find $f(-6)$

17) $g(n) = \frac{16n-8}{6n}$, find $g(2)$

18) $g(n) = \sqrt{5n} - 2$, find $g(5)$

19) $h(x) = x^{-1} - 6$, find $h(\frac{1}{9})$

20) $h(n) = n^{-3} + 4$, find $h(\frac{1}{2})$

21) $h(n) = n^2 - 5$, find $h(\frac{4}{5})$

22) $h(n) = n^3 - 8$, find $h(-\frac{1}{3})$

23) $h(n) = 4n^2 - 42$, find $h(-4)$

24) $h(n) = -5n^2 - 9n$, find $h(7)$

25) $g(n) = \sqrt{4n^2} - \sqrt{5n}$, find $g(5)$

26) $h(a) = \frac{-15a+7}{3a}$, find $h(-b)$

27) $k(a) = 8a - 9$, find $k(a - 3)$

28) $h(x) = \frac{1}{6}x + 7$, find $h(-12x)$

29) $h(x) = 8x^2 + 10$, find $h(\frac{x}{2})$

30) $h(x) = x^4 - 8$, find $h(-2x)$

Adding and Subtracting Functions

✏ **Perform the indicated operation.**

1) $f(x) = 2x + 3$

   $g(x) = x + 4$

   Find $(f - g)(2)$

2) $g(a) = -3a - 8$

   $f(a) = -4a - 12$

   Find $(g - f)(-2)$

3) $h(t) = 7t + 5$

   $g(t) = 3t + 11$

   Find $(h - g)(t)$

4) $g(a) = -5a - 3$

   $f(a) = 3a^2 + 4$

   Find $(g - f)(x)$

5) $g(x) = \frac{2}{7}x - 10$

   $h(x) = \frac{5}{7}x + 10$

   Find $g(14) - h(14)$

6) $h(3) = \sqrt{7x} - 2$

   $g(x) = \sqrt{7x} + 2$

   Find $(h + g)(7)$

7) $f(x) = x^{-3}$

   $g(x) = x^2 + \frac{4}{x}$

   Find $(f - g)(-1)$

8) $h(n) = n^2 + 8$

   $g(n) = -n + 5$

   Find $(h - g)(a)$

9) $g(x) = -2x^2 - 3 - x$

   $f(x) = 7 + x$

   Find $(g - f)(2x)$

10) $g(t) = 4t - 9$

    $f(t) = -t^2 + 5$

    Find $(g + f)(-z)$

11) $f(x) = 3x + 9$

    $g(x) = -4x^2 + 2x$

    Find $(f - g)(-x^2)$

12) $f(x) = -9x^3 - 4x$

    $g(x) = 4x + 12$

    Find $(f + g)(3x^2)$

Multiplying and Dividing Functions

✍ **Perform the indicated operation.**

1) $g(x) = -2x - 5$
   $f(x) = 3x + 4$
   Find $(g.f)(2)$

2) $f(x) = 3x$
   $h(x) = -2x + 5$
   Find $(f.h)(-3)$

3) $g(a) = 5a - 3$
   $h(a) = a - 7$
   Find $(g.h)(-3)$

4) $f(x) = x - 4$
   $h(x) = 4x - 3$
   Find $(\frac{f}{h})(4)$

5) $f(x) = 9a^2$
   $g(x) = 5 + 4a$
   Find $(\frac{f}{g})(3)$

6) $g(a) = \sqrt{5a} + 7$
   $f(a) = (-a)^2 + 3$
   Find $(\frac{g}{f})(5)$

7) $g(t) = t^2 + 5$
   $h(t) = 2t - 5$
   Find $(g.h)(-3)$

8) $g(n) = n^2 + 2n - 4$
   $h(n) = -n + 6$
   Find $(g.h)(1)$

9) $g(a) = (a - 7)^3$
   $f(a) = a^2 + 8$
   Find $(\frac{g}{f})(7)$

10) $g(x) = -x^2 + \frac{4}{5}x + 10$
    $f(x) = x^2 - 3$
    Find $(\frac{g}{f})(5)$

11) $f(x) = x^3 - 3x^2 + 9$
    $g(x) = x - 4$
    Find $(f.g)(x)$

12) $f(x) = 3x - 5$
    $g(x) = x^2 - 4x$
    Find $(f.g)(x^2)$

Composition of Functions

Using $f(x) = 2x - 5$ and $g(x) = -2x$, find:

1) $f(g(0)) =$

2) $f(g(-1)) =$

3) $g(f(1)) =$

4) $g(f(3)) =$

5) $f(g(-2)) =$

6) $g(f(5)) =$

Using $f(x) = -\frac{1}{4}x + \frac{3}{4}$ and $g(x) = x^2$, find:

7) $g(f(4)) =$

8) $g(f(3)) =$

9) $g(g(2)) =$

10) $f(f(1)) =$

11) $g(f(-1)) =$

12) $g(f(7)) =$

Using $f(x) = -5x + 2$ and $g(x) = x + 3$, find:

13) $g(f(0)) =$

14) $f(f(2)) =$

15) $f(g(3)) =$

16) $f(g(-3)) =$

17) $g(f(-5)) =$

18) $f(f(x)) =$

Using $f(x) = \sqrt{x + 9}$ and $g(x) = x - 9$, find:

19) $f(g(9)) =$

20) $g(f(-8)) =$

21) $f(g(18)) =$

22) $f(f(-5)) =$

23) $g(f(7)) =$

24) $g(g(8)) =$

Quadratic Equation

✏ **Multiply.**

1) $(x-2)(x+8) = $ _____

2) $(x+1)(x+9) = $ _____

3) $(x-5)(x+6) = $ _____

4) $(x+7)(x-3) = $ _____

5) $(x-9)(x-8) = $ _____

6) $(4x+2)(x-4) = $ _____

7) $(3x-6)(x+4) = $ _____

8) $(x-9)(2x+7) = $ _____

9) $(5x+3)(x-4) = $ _____

10) $(4x+2)(3x-3) = $ _____

✏ **Factor each expression.**

11) $x^2 - 4x - 21 = $ _____

12) $x^2 + 14x + 45 = $ _____

13) $x^2 - 5x - 24 = $ _____

14) $x^2 - 7x + 6 = $ _____

15) $x^2 + 14x + 33 = $ _____

16) $4x^2 + 38x + 18 = $ _____

17) $5x^2 + 18x - 8 = $ _____

18) $2x^2 + 2x - 40 = $ _____

19) $2x^2 + 22x + 56 = $ _____

20) $12x^2 - 148x + 360 = $ _____

✏ **Calculate each equation.**

21) $(x+6)(x-9) = 0$

22) $(x+1)(x+11) = 0$

23) $(3x+9)(x+3) = 0$

24) $(5x-5)(6x+12) = 0$

25) $x^2 - 12x + 30 = 6$

26) $x^2 + 6x + 14 = 5$

27) $x^2 + \frac{9}{2}x + 7 = 5$

28) $x^2 + 2x - 25 = 10$

29) $2x^2 + 12x - 54 = 0$

30) $x^2 - 11x = 12$

Solving Quadratic Equations

✎ **Solve each equation by factoring or using the quadratic formula.**

1) $(x+5)(x-2) = 0$

2) $(x+8)(x+2) = 0$

3) $(x-9)(x+5) = 0$

4) $(x-3)(x-1) = 0$

5) $(x+9)(x+4) = 0$

6) $(2x+5)(x+9) = 0$

7) $(9x+8)(3x+9) = 0$

8) $(4x+2)(x+5) = 0$

9) $(x+2)(2x+9) = 0$

10) $(12x+3)(2x+9) = 0$

11) $2x^2 = 16x$

12) $x^2 - 16 = 0$

13) $2x^2 + 48 = 22x$

14) $-x^2 - 20 = 9x$

15) $x^2 + 8x = 33$

16) $2x^2 + 12x = 80$

17) $x^2 + 14x = -48$

18) $x^2 + 15x = -54$

19) $x^2 + 15x = -36$

20) $x^2 + 2x - 40 = 5x$

21) $x^2 + 16x = -63$

22) $x^2 - 18x = -81$

23) $10x^2 = 7x - 1$

24) $7x^2 - 5x + 8 = 8$

25) $8x^2 + 27 = 33x$

26) $5x^2 - 26x = -24$

27) $3x^2 + 6 = -19x$

28) $x^2 + 22x = -117$

29) $x^2 + 3x - 58 = 30$

30) $5x^2 + 20x - 200 = 25$

31) $3x^2 - 33x + 84 = 0$

32) $6x^2 - 31x + 30 = 15 - 10x^2$

Quadratic Formula and the Discriminant

✎ **Find the value of the discriminant of each quadratic equation.**

1) $3x(x-9) = 0$

2) $2x^2 + 9x - 4 = 0$

3) $x^2 + 9x + 5 = 0$

4) $4x^2 - 4x + 7 = 0$

5) $x^2 + 7x - 6 = 0$

6) $4x^2 + 5x - 13 = 0$

7) $3x^2 + 7x + 11 = 0$

8) $x^2 - 4x - 12 = 0$

9) $5x^2 + 9x + 8 = 0$

10) $x^2 + 3x - 7 = 0$

11) $6x^2 + 7x - 13 = 0$

12) $-8x^2 - 11x + 9 = 0$

13) $-9x^2 - 13x + 7 = 0$

14) $-6x^2 - 7x - 9 = 0$

15) $14x^2 - 8x - 15 = 0$

16) $-9x^2 - 5x + 10 = 0$

17) $8x^2 + 9x - 14 = 0$

18) $7x^2 - 15x = 0$

19) $3x^2 - 7x + 9 = 0$

20) $7x^2 + 4x + 16 = 0$

✎ **Find the discriminant of each quadratic equation then state the number of real and imaginary solutions.**

21) $-x^2 - 4 = 4x$

22) $20x^2 = 20x - 5$

23) $-11x^2 - 11x = 22$

24) $19x^2 - 4x + 1 = 15x^2$

25) $-8x^2 = -6x + 6$

26) $2x^2 + 4x + 4 = 2$

27) $6x^2 - 2x - 9 = -12$

28) $-14x^2 - 56x - 64 = -8$

Quadratic Inequalities

✏️ **Solve each quadratic inequality.**

1) $x^2 - 64 < 0$

2) $-x^2 - 6x - 8 > 0$

3) $x^2 + 6x + 8 < 0$

4) $4x^2 + 28x + 40 > 0$

5) $5x^2 - 5x - 10 \geq 0$

6) $3x^2 > -12x - 27$

7) $4x^2 + 10x + 28 \leq 0$

8) $3x^2 - 9x - 30 \leq 0$

9) $5x^2 - 35x + 60 \geq 0$

10) $x^2 + 7x + 10 < 0$

11) $2x^2 + 16x - 130 > 0$

12) $8x^2 - 24x + 18 > 0$

13) $2x^2 - 32x + 136 \leq 0$

14) $x^2 - 14x + 49 \leq 0$

15) $2x^2 - 30x + 112 \geq 0$

16) $2x^2 + 16x + 32 \leq 0$

17) $x^2 - 121 \leq 0$

18) $2x^2 - 22x + 60 \geq 0$

19) $8x^2 + 10x + 18 \leq 0$

20) $4x^2 - 2x - 24 > 2x^2$

21) $4x^2 - 16x + 16 < 0$

22) $15x^2 - 6x \geq 14x^2 - 5$

23) $6x^2 - 24 > 4x^2 + 2x$

24) $3x^2 - x \geq 3x^2 - 4x + 6$

25) $2x^2 + 2x - 8 > x^2$

26) $4x^2 + 20x - 11 < 0$

27) $-2x^2 + 30x - 114 \geq 0$

28) $-8x^2 + 6x - 1 \leq 0$

29) $x^2 + 7x + 10 < 0$

30) $36x^2 + 46x + 10 \leq 0$

31) $5x^2 + 5x - 60 \geq 0$

32) $3x^2 + 4x \leq 2x^2 + 2x - 10$

Graphing Quadratic Functions

✎ **Sketch the graph of each function. Identify the vertex and axis of symmetry.**

1) $y = (x + 3)^2 + 5$

2) $y = (x - 3)^2 - 1$

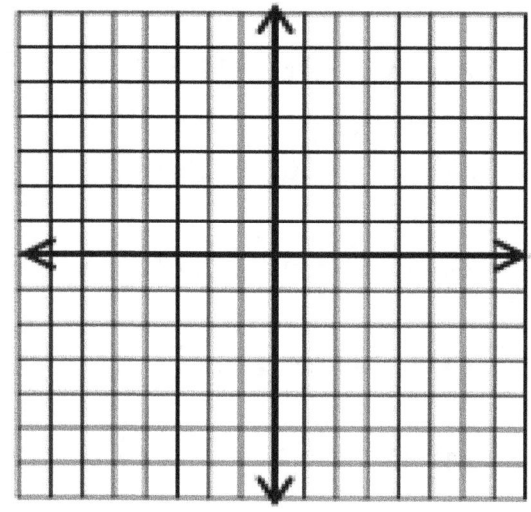 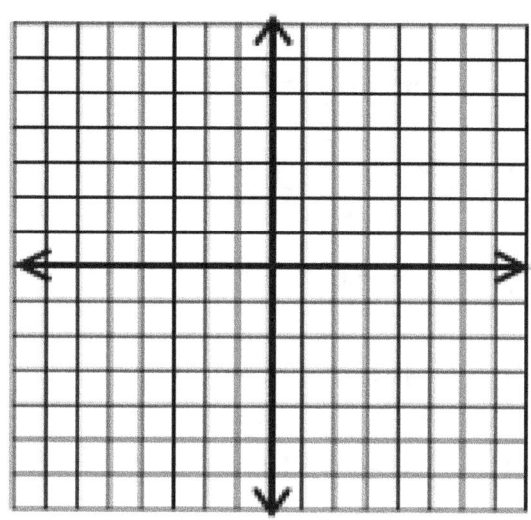

3) $y = 6 - (-x + 2)^2$

4) $y = -3x^2 - 6x + 9$

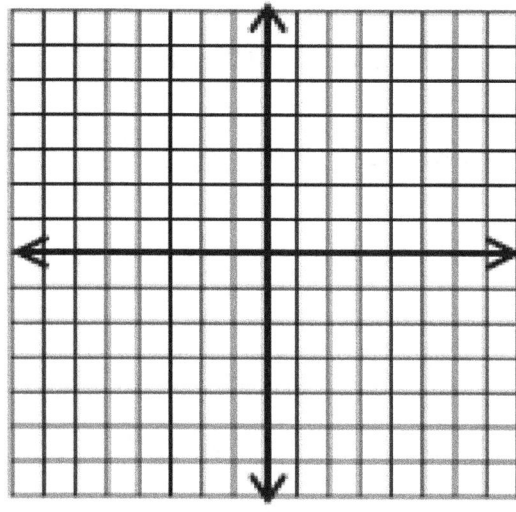 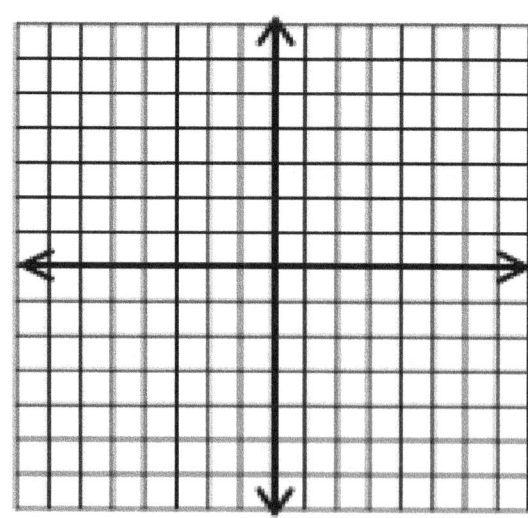

Domain and Range of Radical Functions

✎ **Identify the domain and range of each function.**

1) $y = \sqrt{x+6} - 9$

2) $y = \sqrt[3]{5x-4} - 12$

3) $y = \sqrt{3x-9} + 7$

4) $y = \sqrt[3]{(8x+11)} - 9$

5) $y = 2\sqrt{6x+30} + 14$

6) $y = \sqrt[3]{(9x-15)} - 17$

7) $y = 2\sqrt{9x^2+18} + 7$

8) $y = \sqrt[3]{(8x^2-5)} - 13$

9) $y = \sqrt{2x^3+16} - 9$

10) $y = \sqrt[3]{(14x+3)} - 5x$

11) $y = 3\sqrt{-3(12x+24)} + 7$

12) $y = \sqrt[5]{(11x^2-17)} - 21$

13) $y = 4\sqrt{x-9} - 27$

14) $y = \sqrt[3]{5x+1} - 3$

✎ **Sketch the graph of each function.**

15) $y = -3\sqrt{x} + 4$

16) $y = 6\sqrt{x} - 8$

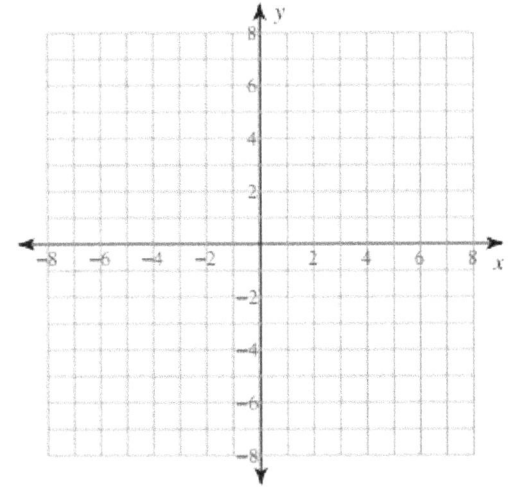
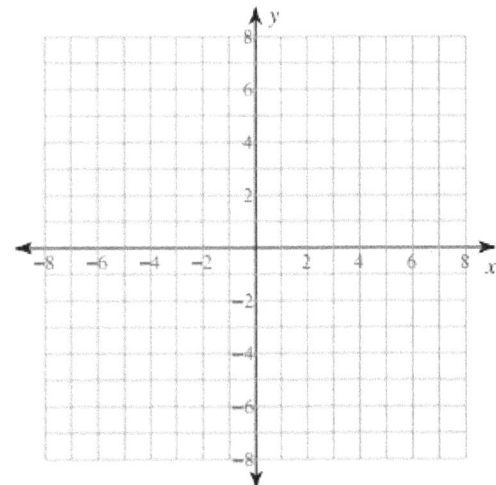

Solving Radical Equations

✎ **Solve each equation. Remember to check for extraneous solutions.**

1) $\sqrt{a} = 9$

2) $\sqrt{v} = 4$

3) $\sqrt{r} = 7$

4) $4 = 16\sqrt{x}$

5) $\sqrt{x+3} = 12$

6) $2 = \sqrt{x-8}$

7) $7 = \sqrt{r-6}$

8) $\sqrt{x-4} = 9$

9) $15 = \sqrt{x-6}$

10) $\sqrt{m+8} = 11$

11) $5\sqrt{3a} = 75$

12) $2\sqrt{10x} = 30$

13) $4 = \sqrt{3x-10}$

14) $\sqrt{150-3x} = 2$

15) $\sqrt{r+4} - 8 = 8$

16) $-18 = -3\sqrt{r+2}$

17) $60 = 6\sqrt{49v}$

18) $3 = \sqrt{50-x}$

19) $\sqrt{90-5a} = 6$

20) $\sqrt{-3n+33} = 3$

21) $\sqrt{21r-18} = 4r$

22) $\sqrt{-14+6x} = 7x$

23) $\sqrt{4x+15} = \sqrt{2x+11}$

24) $\sqrt{8v} = \sqrt{10v-14}$

25) $\sqrt{16-3x} = \sqrt{3x-8}$

26) $\sqrt{5m+12} = \sqrt{7m+12}$

27) $\sqrt{8r+15} = \sqrt{-13-5r}$

28) $\sqrt{4k+6} = \sqrt{2-8k}$

29) $-60\sqrt{x-16} = -120$

30) $\sqrt{20-x} = \sqrt{\dfrac{x}{4}}$

Answers of Worksheets – Chapter 10

Evaluating Function

1) $h(x) = -8x + 9$
2) $k(a) = 5a - 21$
3) $d(t) = 14t$
4) $y(x) = \frac{3}{17}x - \frac{9}{17}$
5) $m(n) = 18n - 94$
6) $c(p) = p^2 - 7p + 15$
7) -25
8) 6.5
9) -1
10) 14
11) -9
12) 38
13) 30
14) -29
15) -3
16) 11.5
17) 2
18) 3
19) 3
20) 12
21) $-\frac{109}{25}$
22) $-\frac{215}{27}$
23) 22
24) -308
25) 5
26) $-\frac{15b+7}{3b}$
27) $8a - 33$
28) $-2x + 7$
29) $2x^2 + 10$
30) $-16x^4 - 8$

Adding and Subtracting Functions

1) 1
2) 2
3) $4t - 6$
4) $-3x^2 - 5x - 7$
5) -26
6) 14
7) 2
8) $a^2 + a + 3$
9) $-8x^2 - 4x - 10$
10) $-z^2 - 4z - 4$
11) $4x^4 - x^2 + 9$
12) $-243x^6 + 12$

Multiplying and Dividing Functions

1) -90
2) -99
3) 180
4) 0
5) $\frac{81}{17}$
6) $\frac{3}{7}$
7) -154
8) -5
9) 0
10) $-\frac{1}{2}$
11) $x^4 - 7x^3 + 12x^2 + 9x - 36$
12) $3x^6 - 17x^4 + 20x^2$

Composition of Functions

1) -5
2) -1
3) 6
4) -2

5) 3

6) −10

7) $\frac{1}{16}$

8) 0

9) 16

10) $\frac{5}{8}$

11) 1

12) 1

13) 5

14) 42

15) −28

16) 2

17) 30

18) $25x − 8$

19) 3

20) −8

21) $3\sqrt{2}$

22) $\sqrt{11}$

23) −5

24) −10

Quadratic Equations

1) $x^2 + 6x − 16$

2) $x^2 + 10x + 9$

3) $x^2 + x − 30$

4) $x^2 + 4x − 21$

5) $x^2 − 17x + 72$

6) $4x^2 − 14x − 8$

7) $3x^2 + 6x − 24$

8) $2x^2 − 11x − 63$

9) $5x^2 − 17x − 12$

10) $12x^2 − 6x − 6$

11) $(x − 7)(x + 3)$

12) $(x + 5)(x + 9)$

13) $(x − 8)(x + 3)$

14) $(x − 1)(x − 6)$

15) $(x + 3)(x + 11)$

16) $(4x + 2)(x + 9)$

17) $(5x − 2)(x + 4)$

18) $(2x − 8)(x + 5)$

19) $(2x + 8)(x + 7)$

20) $4(x − 9)(3x − 10)$

21) $x = −6, x = 9$

22) $x = −1, x = −11$

23) $x = −3$

24) $x = 1, x = −2$

25) $x = 6$

26) $x = −3$

27) $x = −4, x = −\frac{1}{2}$

28) $x = 5, x = −7$

29) $x = 3, x = −9$

30) $x = −1, x = 12$

Solving quadratic equations

1) $\{−5, 2\}$

2) $\{−8, −2\}$

3) $\{9, −5\}$

4) $\{3, 1\}$

5) $\{−9, −4\}$

6) $\{−\frac{5}{2}, −9\}$

7) $\{−\frac{8}{9}, −3\}$

8) $\{−\frac{1}{2}, −5\}$

9) $\{−2, −\frac{9}{2}\}$

10) $\{−\frac{1}{4}, −\frac{9}{2}\}$

11) $\{8, 0\}$

12) $\{4, −4\}$

13) $\{3, 8\}$

14) $\{−5, −4\}$

15) $\{3, −11\}$

16) $\{4, −10\}$

17) $\{−6, −8\}$

18) $\{−6, −9\}$

19) $\{−3, −12\}$

20) $\{8, −5\}$

21) $\{−7, −9\}$

22) $\{9\}$

23) $\{\frac{1}{5}, \frac{1}{2}\}$

24) $\{\frac{5}{7}, 0\}$

25) $\{\frac{9}{8}, 3\}$

26) $\{\frac{6}{5}, 4\}$

27) $\{−\frac{1}{3}, −6\}$

28) $\{−9, −13\}$

29) $\{8, −11\}$

30) $\{5, −9\}$

31) $\{4, 7\}$

32) $\{\frac{15}{16}, 1\}$

Quadratic formula and the discriminant

1) 729
2) 113
3) 61
4) −96
5) 73
6) 233
7) 83
8) 8
9) −79
10) 37
11) 361
12) 409
13) 421
14) 7
15) 904
16) 385
17) 529
18) 225
19) −59
20) −432
21) 0, one real solution
22) 0, one real solution
23) −847, no solution
24) 0, one real solution
25) −156, no solution
26) 0, one real solution
27) −68, no solution
28) 0, one real solution

Solve quadratic inequalities

1) $-8 < x < 8$
2) $-4 < x < -2$
3) $-4 < x < -2$
4) $x < -5$ or $x > -2$
5) $x \leq -1$ or $x \geq 2$
6) all real numbers
7) no solution
8) $-2 \leq x \leq 5$
9) $x \leq 3$ or $x \geq 4$
10) $-5 < x < -2$
11) $x < -13$ or $x > 5$
12) $x < \frac{3}{2}$ or $x > \frac{3}{2}$
13) no solution
14) $x = 7$
15) $x \leq 7$ or $x \geq 8$
16) $x = -4$
17) $-11 \leq x \leq 11$
18) $x \leq 5$ or $x \geq 6$
19) no solution
20) $x < -3$ or $x > 4$
21) no solution
22) $x \leq 1$ or $x \geq 5$
23) $x < -3$ or $x > 4$
24) $x \geq 2$
25) $x < -4$ or $x > 2$
26) $-\frac{11}{2} < x < \frac{1}{2}$
27) no solution
28) $x \leq \frac{1}{4}$ or $x \geq \frac{1}{2}$
29) $-5 < x < -2$
30) $-1 \leq x \leq -\frac{5}{18}$
31) $x \leq -4$ or $x \geq 3$
32) no solution

Graphing quadratic functions

1) $(-3, 5), x = -3$

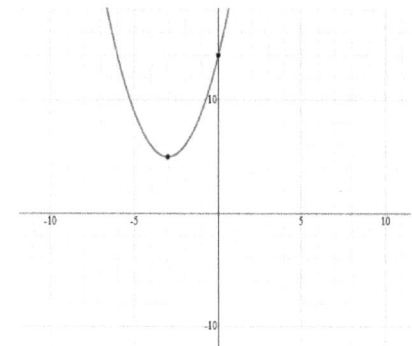

2) $(3, -1), x = 3$

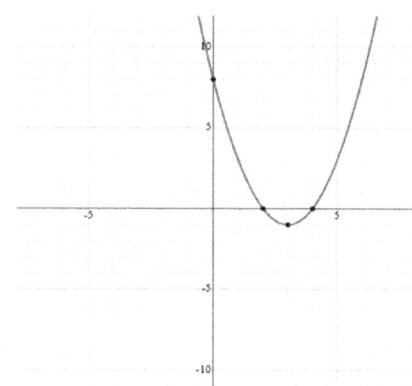

3) $(2, 6), x = 2$

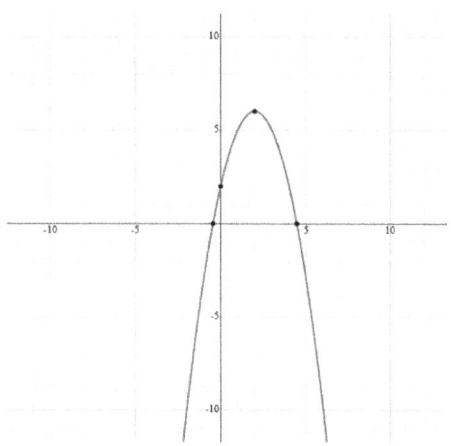

1) $(-1, 12), x = -1$

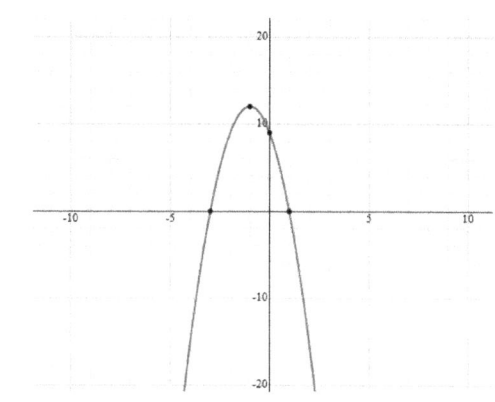

Domain and range of radical functions

1) domain: $x \geq -6$
   range: $y \geq -9$

2) domain: {all real numbers}
   range: {all real numbers}

3) domain: $x \geq 3$
   range: $y \geq 7$

4) domain: {all real numbers}
   range: {all real numbers}

5) domain: $x \geq -5$
   range: $y \geq 14$

6) domain: {all real numbers}
   range: {all real numbers}

7) domain: {all real numbers}
   range: $y \geq 6\sqrt{2} + 7$

8) domain: {all real numbers}
   range: {all real numbers}

9) domain: $x \geq -2$
   range: $y \geq -9$
10) domain: {all real numbers}
    range: {all real numbers}
11) domain: $x \leq -2$
    range: $y \geq 7$
12) domain: {all real numbers}
    range: {all real numbers}
13) domain: $x \geq 9$
    range: $y \geq -27$
14) domain: {all real numbers}
    range: {all real numbers}

15)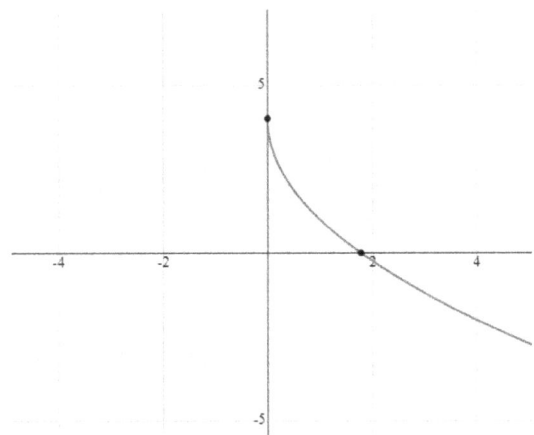

16)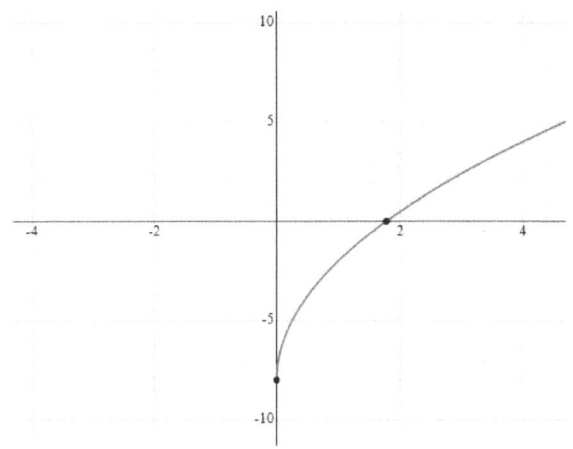

Solving radical equations

1) {81}
2) {16}
3) {49}
4) {$\frac{1}{16}$}
5) {141}
6) {12}
7) {55}
8) {85}
9) {231}
10) {113}
11) {75}
12) {22.5}
13) {$\frac{26}{3}$}
14) $\frac{146}{3}$
15) {252}
16) {34}
17) {$\frac{100}{49}$}
18) {41}
19) {54/5}
20) {8}
21) no solution
22) no solution
23) {−2}
24) {7}
25) {4}
26) {0}
27) no solution
28) {$-\frac{1}{3}$}
29) {20}
30) {16}

Chapter 11:
Complex Numbers

Topics that you will practice in this chapter:

- ✓ Adding and Subtracting Complex Numbers
- ✓ Multiplying and Dividing Complex Numbers
- ✓ Graphing Complex Numbers
- ✓ Rationalizing Imaginary Denominators

Mathematics is a hard thing to love. It has the unfortunate habit, like a rude dog, of turning its most unfavorable side towards you when you first make contact with it. — David Whiteland

Adding and Subtracting Complex Numbers

✎ **Simplify.**

1) $(8i) - (4i) =$

2) $(5i) + (2i) =$

3) $(2i) + (7i) =$

4) $(-6i) - (i) =$

5) $(12i) + (4i) =$

6) $(4i) - (-4i) =$

7) $(-4i) + (-5i) =$

8) $(13i) - (6i) =$

9) $(-21i) - (7i) =$

10) $(-4i) + (2 + 8i) =$

11) $(8 - 4i) + (-6i) =$

12) $(-3i) + (9 + 12i) =$

13) $5 + (9 - 2i) =$

14) $(10i) - (-6 + 2i) =$

15) $(3 + 9i) - (-4i) =$

16) $(7 + 8i) + (-5i) =$

17) $(5i) - (-3i + 4) =$

18) $(6i + 2) + (-2i) =$

19) $(12) - (18 + 3i) =$

20) $(7 + 3i) + (6 + 2i) =$

21) $(4 - 9i) + (3 + 8i) =$

22) $(7 + 3i) + (10 + 12i) =$

23) $(-5 + 5i) - (-5 - 7i) =$

24) $(-8 + 12i) - (-9 + 8i) =$

25) $(-18 + 3i) - (-3 - 12i) =$

26) $(-13 - 4i) + (9 + 12i) =$

27) $(-15 - 2i) - (-14 - 6i) =$

28) $-4 + (8i) + (-14 + 7i) =$

29) $19 - (8i) + (2 - 5i) =$

30) $-3 + (-4 - 8i) - 9 =$

31) $(-24i) + (5 - 8i) + 12 =$

32) $(-11i) - (15 - 12i) + 9i =$

Multiplying and Dividing Complex Numbers

✎ **Simplify.**

1) $(5i)(-3i) =$

2) $(-7i)(2i) =$

3) $(3i)(3i)(-3i) =$

4) $(6i)(-6i) =$

5) $(-7-6i)(7+6i) =$

6) $(4-2i)^2 =$

7) $(5-2i)(4-2i) =$

8) $(5+2i)^2 =$

9) $(7i)(-3i)(9-2i) =$

10) $(2-8i)(6-8i) =$

11) $(-9+3i)(1+4i) =$

12) $(7-8i)(9-3i) =$

13) $5(3i) - (5i)(-4+2i) =$

14) $\dfrac{5}{-25i} =$

15) $\dfrac{12-9i}{-3i} =$

16) $\dfrac{4+9i}{i} =$

17) $\dfrac{20i}{-6+2i} =$

18) $\dfrac{-4-11i}{2i} =$

19) $\dfrac{7i}{3-i} =$

20) $\dfrac{4-9i}{12-5i} =$

21) $\dfrac{8-3i}{-4-4i} =$

22) $\dfrac{-9-5i}{-3-i} =$

23) $\dfrac{-4+i}{-6-5i} =$

24) $\dfrac{-6-7i}{-3+4i} =$

25) $\dfrac{8+11i}{5-5i} =$

Graphing Complex Numbers

✎ **Identify each complex number graphed.**

1)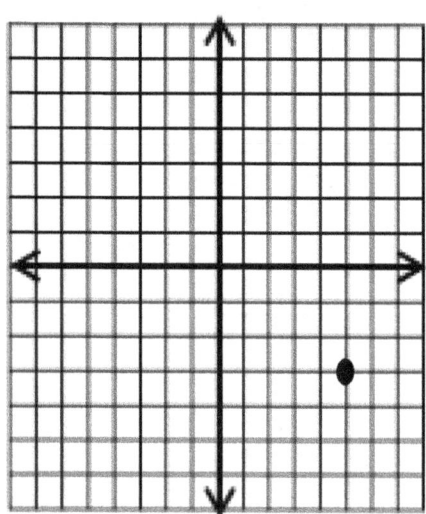

2)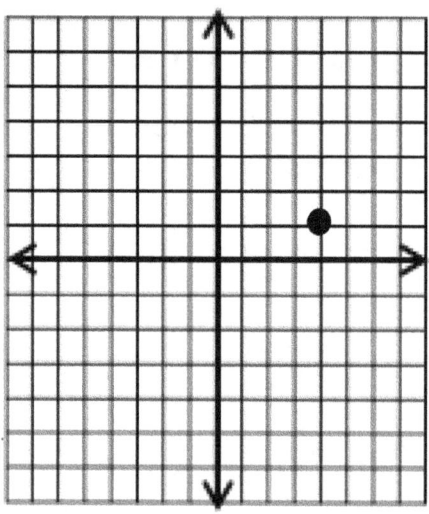

3)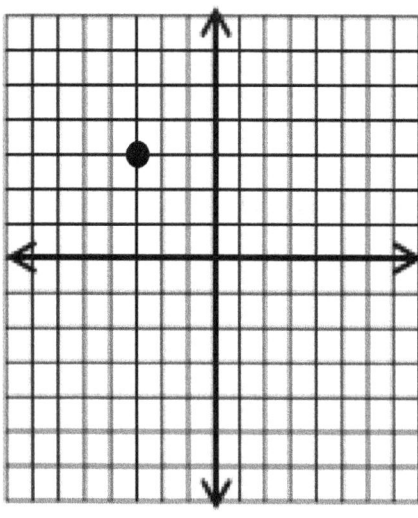

4)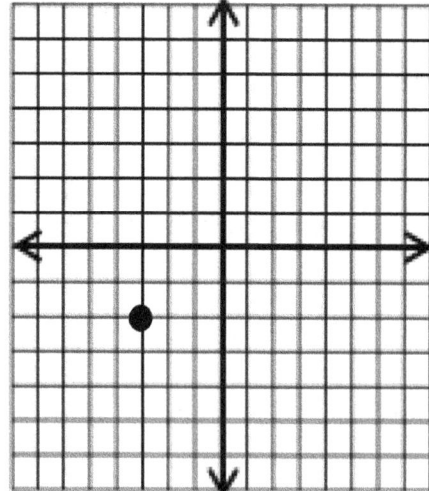

Rationalizing Imaginary Denominators

✎ **Simplify.**

1) $\dfrac{-8}{-8i} =$

2) $\dfrac{-3}{-21i} =$

3) $\dfrac{-14}{-28i} =$

4) $\dfrac{-30}{-5i} =$

5) $\dfrac{9}{2i} =$

6) $\dfrac{27}{-9i} =$

7) $\dfrac{45}{-20i} =$

8) $\dfrac{-26}{8i} =$

9) $\dfrac{6x}{3yi} =$

10) $\dfrac{9-9i}{-3i} =$

11) $\dfrac{4-9i}{-i} =$

12) $\dfrac{12+4i}{3i} =$

13) $\dfrac{8i}{-1+4i} =$

14) $\dfrac{8i}{-2+6i} =$

15) $\dfrac{-15-3i}{-4+4i} =$

16) $\dfrac{-5-9i}{3+4i} =$

17) $\dfrac{-11-4i}{5-2i} =$

18) $\dfrac{-4+6i}{-3i} =$

19) $\dfrac{9+5i}{4i} =$

20) $\dfrac{-5-3i}{7-2i} =$

21) $\dfrac{-8+i}{-3i} =$

22) $\dfrac{9+i}{-5-2i} =$

23) $\dfrac{-9-5i}{-7-2i} =$

24) $\dfrac{9i-5}{-3-6i} =$

CLEP College Math Workbook

Answers of Worksheets – Chapter 11

Adding and Subtracting Complex Numbers

1) $4i$
2) $7i$
3) $9i$
4) $-7i$
5) $16i$
6) $8i$
7) $-9i$
8) $7i$
9) $-28i$
10) $2 + 4i$
11) $8 - 10i$
12) $9 + 9i$
13) $14 - 2i$
14) $6 + 8i$
15) $3 + 13i$
16) $7 + 3i$
17) $-4 + 8i$
18) $2 + 4i$
19) $-6 - 3i$
20) $13 + 5i$
21) $7 - i$
22) $17 + 15i$
23) $12i$
24) $1 + 4i$
25) $-15 + 15i$
26) $-4 + 8i$
27) $-1 + 4i$
28) $-18 + 15i$
29) $21 - 13i$
30) $-16 - 8i$
31) $17 - 32i$
32) $-15 + 10i$

Multiplying and Dividing Complex Numbers

1) 15
2) 14
3) $27i$
4) 36
5) $-13 - 84i$
6) $-16i + 12$
7) $16 - 18i$
8) $21 + 20i$
9) $189 - 42i$
10) $-52 - 64i$
11) $-21 - 33i$
12) $39 - 93i$
13) $10 + 35i$
14) $\frac{i}{5}$
15) $3 + 4i$
16) $9 - 4i$
17) $1 - 3i$
18) $\frac{11}{2} - 2i$
19) $-\frac{7}{10} + \frac{21}{10}i$
20) $\frac{93}{169} - \frac{88}{169}i$
21) $-\frac{5}{8} + \frac{11}{8}i$
22) $\frac{16}{5} + \frac{3}{5}i$
23) $\frac{19}{61} - \frac{26}{61}i$
24) $-\frac{2}{5} + \frac{9}{5}i$
25) $-\frac{3}{10} + \frac{19}{10}i$

Graphing Complex Numbers

2) $5 - 3i$
3) $4 + i$
4) $-3 + 3i$
5) $-3 - 2i$

Rationalizing Imaginary Denominators

1) $-i$
2) $-\frac{1}{7}i$
3) $\frac{-1}{2}i$
4) $-6i$
5) $-\frac{9}{2}i$
6) $3i$
7) $\frac{9}{4}i$
8) $\frac{13}{4}i$
9) $-\frac{2x}{y}i$
10) $3 + 3i$
11) $9 + 4i$
12) $-\frac{4}{3} + 4i$
13) $\frac{32}{17} - \frac{8}{17}i$
14) $\frac{6}{5} - \frac{2}{5}i$
15) $\frac{3}{2} + \frac{9}{4}i$
16) $-\frac{51}{25} - \frac{7}{25}i$
17) $-\frac{47}{29} - \frac{42}{29}i$
18) $-2 - \frac{4}{3}i$
19) $-\frac{5}{4} + \frac{9}{4}i$
20) $-\frac{29}{53} - \frac{31}{53}i$
21) $-\frac{1}{3} - \frac{8}{3}i$
22) $-\frac{47}{29} + \frac{13}{29}i$
23) $\frac{73}{53} + \frac{17}{53}i$
24) $-\frac{13}{15} - \frac{19}{15}i$

WWW.MathNotion.Com

Chapter 12: Logarithms

Topics that you will practice in this chapter:

- ✓ Rewriting Logarithms
- ✓ Evaluating Logarithms
- ✓ Properties of Logarithms
- ✓ Natural Logarithms
- ✓ Exponential Equations Requiring Logarithms
- ✓ Solving Logarithmic Equations

Mathematics is an art of human understanding. — *William Thurston*

Rewriting Logarithms

✍ **Rewrite each equation in exponential form.**

1) $\log_5 25 = 2$

2) $\log_4 256 = 4$

3) $\log_3 81 = 4$

4) $\log_8 64 = 2$

5) $\log_6 216 = 3$

6) $\log_2 16 = 4$

7) $\log_{10} 100 = 2$

8) $\log_3 243 = 5$

9) $\log_5 625 = 4$

10) $\log_2 256 = 8$

11) $\log_3 6,561 = 8$

12) $\log_{11} 121 = 2$

13) $\log_{14} 196 = 2$

14) $\log_{81} 3 = \frac{1}{4}$

15) $\log_{27} 3 = \frac{1}{3}$

16) $\log_{32} 2 = \frac{1}{5}$

17) $\log_{512} 8 = \frac{1}{3}$

18) $\log_2 \frac{1}{8} = -3$

19) $\log_2 \frac{1}{16} = -4$

20) $\log_a \frac{7}{3} = b$

✍ **Rewrite each exponential equation in logarithmic form.**

21) $12^2 = 144$

22) $7^3 = 343$

23) $4^5 = 1,024$

24) $15^2 = 225$

25) $5^4 = 625$

26) $6^4 = 1,296$

27) $2^9 = 512$

28) $5^5 = 3,125$

29) $4^{-6} = \frac{1}{4,096}$

30) $3^{-5} = \frac{1}{243}$

31) $16^{-2} = \frac{1}{256}$

32) $6^{-3} = \frac{1}{216}$

33) $3^{-9} = \frac{1}{19,683}$

34) $21^{-2} = \frac{1}{441}$

Evaluating Logarithms

✎ **Evaluate each logarithm.**

1) $\log_3 2,187 =$

2) $\log_2 256 =$

3) $\log_5 125 =$

4) $\log_5 625 =$

5) $\log_3 243 =$

6) $\log_4 1,024 =$

7) $\log_8 64 =$

8) $\log_8 \frac{1}{8} =$

9) $\log_6 \frac{1}{36} =$

10) $\log_2 \frac{1}{16} =$

11) $\log_6 \frac{1}{216} =$

12) $\log_3 \frac{1}{256} =$

13) $\log_{18} \frac{1}{324} =$

14) $\log_{256} \frac{1}{4} =$

15) $\log_{512} 8 =$

16) $\log_4 \frac{1}{4,096} =$

17) $\log_9 \frac{1}{729} =$

18) $\log_{216} \frac{1}{6} =$

✎ **Circle the points which are on the graph of the given logarithmic functions.**

19) $y = 5\log_8(3x - 4) + 1$ $(6, 5)$, $(4, 6)$, $(4, 8)$

20) $y = 3\log_2(4x) - 6$ $(4, 6)$, $(\frac{1}{4}, 16)$, $(\frac{1}{4}, -6)$

21) $y = -3\log_5(x - 2) + 5$ $(7, -2)$, $(7, 2)$, $(6, -3)$

22) $y = \frac{1}{2}\log_6(6x) + 4$ $(6, 5)$, $(6, \frac{1}{5})$, $(6, -5)$

23) $y = -\log_9 9(x + 5) + 4$ $(-4, 2)$, $(4, 0)$, $(4, 2)$

24) $y = -\log_8(x - 6) - 4$ $(7, -\frac{1}{4})$, $(7, -4)$, $(7, -\frac{1}{4})$

25) $y = -3\log_6(x + 3) + 6$ $(3, 3)$, $(-5, -3)$, $(-5, 3)$

Properties of Logarithms

✎ **Expand each logarithm.**

1) $\log(9 \times 4) =$

2) $\log(6 \times 3) =$

3) $\log(2 \times 8) =$

4) $\log\left(\frac{8}{7}\right) =$

5) $\log\left(\frac{9}{5}\right) =$

6) $\log\left(\frac{4}{11}\right)^3 =$

7) $\log(9 \times 4^3) =$

8) $\log\left(\frac{7}{3}\right)^2 =$

9) $\log\left(\frac{5^4}{9}\right) =$

10) $\log(x \times y)^7 =$

11) $\log(x^2 \times y \times z^5) =$

12) $\log\left(\frac{u^8}{v}\right) =$

13) $\log\left(\frac{x}{y^4}\right) =$

✎ **Condense each expression to a single logarithm.**

14) $\log 7 - \log 12 =$

15) $\log 8 + \log 3 =$

16) $4 \log 2 - 7 \log 5 =$

17) $6 \log 4 - 9 \log 5 =$

18) $3 \log 8 - \log 17 =$

19) $8 \log 3 - 6 \log 2 =$

20) $\log 11 - 2 \log 5 =$

21) $4 \log 6 + 3 \log 9 =$

22) $4 \log 5 + 5 \log 13 =$

23) $7 \log_5 a + 16 \log_5 b =$

24) $5 \log_6 x - 7 \log_6 y =$

25) $\log_5 u - 9 \log_5 v =$

26) $8 \log_3 u + 21 \log_3 v =$

27) $26 \log_7 u - 15 \log_7 v =$

Natural Logarithms

📝 **Solve each equation for x.**

1) $e^x = 9$
2) $e^x = 36$
3) $e^x = 49$
4) $\ln x = 3$
5) $\ln(\ln x) = 3$
6) $e^x = 11$
7) $\ln(5x + 9) = 1$
8) $\ln(7x + 3) = 3$
9) $\ln(8x + 5) = 4$
10) $\ln x = \frac{1}{3}$
11) $\ln 6x = e^4$
12) $\ln x = \ln 4 + \ln 7$
13) $\ln x = 3\ln 3 + \ln 8$

📝 **Evaluate without using a calculator.**

14) $3\ln e =$
15) $\ln e^{10} =$
16) $4 \ln e =$
17) $\ln e^{15} =$
18) $13\ln e =$
19) $3\ln e^4 =$
20) $e^{\ln 19} =$
21) $e^{2\ln 5} =$
22) $e^{4\ln 3} =$
23) $\ln \sqrt[6]{e} =$

📝 **Reduce the following expressions to simplest form.**

24) $e^{-2\ln 6 + 2\ln 4} =$
25) $e^{-2\ln\left(\frac{4}{5e}\right)} =$
26) $2\ln(e^6) =$
27) $\ln\left(\frac{1}{e}\right)^9 =$
28) $e^{\ln 6 + 3\ln 5} =$
29) $e^{\ln\left(\frac{13}{e}\right)} =$
30) $7\ln(1^{-3e}) =$
31) $\ln\left(\frac{1}{e}\right)^{-12} =$
32) $3\ln\left(\frac{\sqrt[6]{e}}{3e}\right) =$
33) $e^{-3\ln e + 3\ln 3} =$
34) $e^{\ln\frac{15}{e}} =$
35) $19\ln(e^e) =$

Exponential Equations and Logarithms

✎ **Solve each equation for the unknown variable.**

1) $3^{2n} = 27$

2) $5^r = 125$

3) $15^n = 85$

4) $8^{r+3} = 2$

5) $144^x = 12$

6) $7^{-3v-2} = 49$

7) $8^{2n} = 64$

8) $6^n = 1{,}296$

9) $\dfrac{15^{2a}}{3^{-a}} = 315$

10) $11 \times 11^{-v} = 1{,}331$

11) $3^{2n} = \dfrac{1}{81}$

12) $\left(\dfrac{1}{9}\right)^n = 81$

13) $256^{2x} = 4$

14) $9^{3-2x} = 9^{-x}$

15) $6^{-3x} = 6^{x-3}$

16) $2^{3n} = 32$

17) $12^{5x+3} = 12^{2x}$

18) $10^{2n} = 100$

19) $3^{-4k} = 243$

20) $3^r = 9^{-4r}$

21) $13^{x+3} = 13^{4x}$

22) $9^{3x} = 729$

23) $15 \times 15^{-v} = 225$

24) $\dfrac{81}{3^{-2m}} = 3^{-2m-1}$

25) $8^{-2n} \times 8^2 = 8^{-n}$

26) $\left(\dfrac{1}{9}\right)^{2n+1} \times \left(\dfrac{1}{9}\right)^{-n-10} = \left(\dfrac{1}{9}\right)^{-2n}$

✎ **Solve each problem. (Round to the nearest whole number)**

27) A substance decays 15% each day. After 11 days, there are 6 milligrams of the substance remaining. How many milligrams were there initially? _____

28) A culture of bacteria grows continuously. The culture doubles every 4 hours. If the initial number of bacteria is 13, how many bacteria will there be in 23 hours? _____

29) Bob plans to invest $12,000 at an annual rate of 6.5%. How much will Bob have in the account after six years if the balance is compounded quarterly? _____

30) Suppose you plan to invest $8,000 at an annual rate of 6%. How much will you have in the account after 4 years if the balance is compounded monthly? _____

Solving Logarithmic Equations

✎ **Find the value of the variables in each equation.**

1) $\log(x) + 8 = 4$

2) $-\log_3 4x = 5$

3) $\log(x) + 7 = 6$

4) $\log x - \log 7 = 4$

5) $\log x + \log 4 = 2$

6) $\log 4 + \log x = 3$

7) $\log x + \log 2 = \log 12$

8) $-3\log_3(x - 2) = -15$

9) $\log 4x = \log (3x + 2)$

10) $\log (2k - 4) = \log (k - 5)$

11) $\log(5p - 2) = \log(-2p + 12)$

12) $-8 + \log_3 (n + 3) = -8$

13) $\log_3(x + 5) = \log_3 (x^2 + 8)$

14) $\log_9 (v^2 + 24) = \log_9 (-3v - 8)$

15) $\log (9 + 4b) = \log (7b^2 + 6b)$

16) $\log_9(x + 8) - \log_9 x = \log_9 7$

17) $\log_5 9 + \log_5 x^2 = \log_5 81$

18) $\log_6(x + 5) + \log_6 x = \log_6 24$

✎ **Find the value of x in each natural logarithm equation.**

19) $\ln 9 - \ln(3x + 9) = 3$

20) $\ln(x - 4) - \ln(x - 3) = \ln 4$

21) $\ln e^{27} - \ln(x + 3) = 3$

22) $\ln(2x - 6) - \ln(x - 12) = \ln 10$

23) $\ln 6x + \ln(x - 2) = \ln 3x$

24) $\ln(x - 3) - 2\ln(x - 3) = \ln 9$

25) $\ln (9x + 3) - \ln 5 = 6$

26) $\ln(x - 5) + \ln(x - 4) = \ln 2$

27) $\ln 8 + \ln(x + 4) = 10$

28)

29) $3 \ln 3x - \ln(x + 9) = 3 \ln 3x$

30) $\ln x^2 + \ln x^4 = \ln 1$

31) $\ln x^6 - \ln(x + 6) = 6 \ln x$

32) $16 \ln(x - 2) = 4 \ln(x^2 - 4x + 4)$

33) $\ln(x^2 + 10) = \ln(3x + 8)$

34) $6 \ln x - 6\ln(x + 3) = 12\ln(x^2)$

35) $\ln(2x - 3) - \ln(4x - 3) = \ln 4$

36) $\ln 3 + 9 \ln(x + 2) = \ln 3$

37) $3\ln e^2 + \ln(3x - 2) = \ln 3 + 9$

Answers of Worksheets – Chapter 12

Rewriting Logarithms

1) $5^2 = 25$
2) $4^4 = 256$
3) $3^4 = 81$
4) $8^2 = 64$
5) $6^3 = 216$
6) $2^4 = 16$
7) $10^2 = 100$
8) $3^5 = 243$
9) $5^4 = 625$
10) $2^8 = 256$
11) $3^8 = 6,561$
12) $11^2 = 121$
13) $14^2 = 196$
14) $81^{\frac{1}{4}} = 3$
15) $27^{\frac{1}{3}} = 3$
16) $32^{\frac{1}{5}} = 2$
17) $512^{\frac{1}{3}} = 8$
18) $2^{-3} = \frac{1}{8}$
19) $2^{-4} = \frac{1}{16}$
20) $a^b = \frac{7}{3}$
21) $\log_{12} 144 = 2$
22) $\log_3 343 = 7$
23) $\log_4 1,024 = 5$
24) $\log_{15} 225 = 2$
25) $\log_5 625 = 4$
26) $\log_6 1,296 = 4$
27) $\log_2 512 = 9$
28) $\log_5 3,125 = 5$
29) $\log_4 \frac{1}{4,096} = -6$
30) $\log_3 \frac{1}{243} = -5$
31) $\log_{16} \frac{1}{256} = -2$
32) $\log_6 \frac{1}{216} = -3$
33) $\log_3 \frac{1}{19,683} = -9$
34) $\log_{21} \frac{1}{441} = -2$

Evaluating Logarithms

1) 7
2) 8
3) 3
4) 4
5) 5
6) 5
7) 2
8) −1
9) −2
10) −4
11) −3
12) −4
13) −2
14) $-\frac{1}{4}$
15) $\frac{1}{3}$
16) −6
17) −3
18) $-\frac{1}{3}$
19) (4, 6)
20) $(\frac{1}{4}, -6)$
21) (7, 2)
22) (6, 5)
23) (4, 2)
24) (7, −4)
25) (3, 3)

Properties of Logarithms

1) log 9 + log 4
2) log 6 + log 3
3) log 2 + log 8
4) log 8 − log 7
5) log 9 − log 5
6) 3 log 4 − 3 log 11
7) log 9 + 3 log 4
8) 2 log 7 − 2 log 3

9) $4 \log 5 - \log 9$

10) $7 \log x + 7 \log y$

11) $2\log x + \log y + 5 \log z$

12) $8 \log u - \log v$

13) $\log x - 4 \log y$

14) $\log \frac{7}{12}$

15) $\log(8 \times 3)$

16) $\log \frac{2^4}{5^7}$

17) $\log \frac{4^6}{9^5}$

18) $\log \frac{8^3}{17}$

19) $\log \frac{3^8}{2^6}$

20) $\log \frac{11}{5^2}$

21) $\log (6^4 \times 9^3)$

22) $\log (5^4 \times 13^5)$

23) $\log_5 (a^7 b^{16})$

24) $\log_6 \frac{x^5}{y^7}$

25) $\log_5 \frac{u}{v^9}$

26) $\log_3 (u^8 \times v^{21})$

27) $\log_7 \frac{u^{26}}{v^{15}}$

Natural Logarithms

1) $x = \ln 9$
2) $x = \ln 36, x = 2\ln (6)$
3) $x = \ln 49, x = 2\ln (7)$
4) $x = e^3$
5) $x = e^{e^3}$
6) $x = \ln 11$
7) $x = \frac{e-9}{5}$
8) $x = \frac{e^3 - 3}{7}$
9) $x = \frac{e^4 - 5}{8}$
10) $x = \sqrt[3]{e}$
11) $x = \frac{e^{e^4}}{6}$
12) $x = 28$
13) $x = 216$
14) 3
15) 10
16) 4
17) 15
18) 13
19) 12
20) 19
21) 25
22) 81
23) $\frac{1}{6}$
24) $\frac{4}{9}$
25) $\frac{25}{16e^2}$
26) 12
27) -9
28) 750
29) $\frac{13}{e}$
30) 0
31) 12
32) -5.8
33) $27e^{-3} = \frac{27}{e^3}$
34) $\frac{15}{e}$
35) $19e$

Exponential Equations and Logarithms

1) $\frac{3}{2}$
2) 3
3) 1.64
4) $\frac{-8}{3}$
5) $\frac{1}{2}$
6) $-\frac{4}{3}$
7) 1
8) 0.883

9) -2
10) -2
11) -2
12) $\frac{1}{8}$
13) 3
14) $\frac{3}{4}$
15) $\frac{5}{3}$

16) -1
17) 1
18) $-\frac{5}{4}$
19) 0
20) 1
21) 1
22) -1
23) -1.25

24) 2
25) 3
26) 35.9
27) 699.6
28) $\$17,668.3$
29) $\$10,163.9$

Solving Logarithmic Equations

1) $\{\frac{1}{10,000}\}$
2) $\{\frac{1}{972}\}$
3) $\{\frac{1}{10}\}$
4) $\{70,000\}$
5) $\{25\}$
6) $\{250\}$
7) $\{6\}$
8) $\{245\}$
9) $\{2\}$
10) No Solution
11) $\{2\}$
12) $\{-2\}$
13) No Solution

14) No Solution
15) $\{1, -\frac{9}{7}\}$
16) $\{\frac{4}{3}\}$
17) $\{3, -3\}$
18) $\{3\}$
19) $x = \frac{3-3e^3}{e^3}$
20) No Solution
21) $e^{24} - 3$
22) $\{\frac{57}{4}\}$
23) $\{\frac{5}{2}\}$
24) $\{\frac{28}{9}\}$

25) $x = \frac{5e^6 - 3}{9}$
26) $x = 6$
27) $x = \frac{e^{10} - 32}{8}$
28) No Solution
29) $\{1, -1\}$
30) No Solution
31) $x = 3$
32) $\{1, 2\}$
33) $\{0.64951 ...\}$
34) No Solution
35) $\{-1\}$
36) $x = \frac{3e^3 + 2}{3}$

CLEP College Math Workbook

Chapter 13:
Geometry and Solid Figures

Topics that you will practice in this chapter:

- ✓ Angles
- ✓ Pythagorean Relationship
- ✓ Triangles
- ✓ Polygons
- ✓ Trapezoids
- ✓ Circles
- ✓ Cubes
- ✓ Rectangular Prism
- ✓ Cylinder
- ✓ Pyramids and Cone

Mathematics is, as it were, a sensuous logic, and relates to philosophy as do the arts, music, and plastic art to poetry. — K. Shegel

Angles

✏️ **What is the value of x in the following figures?**

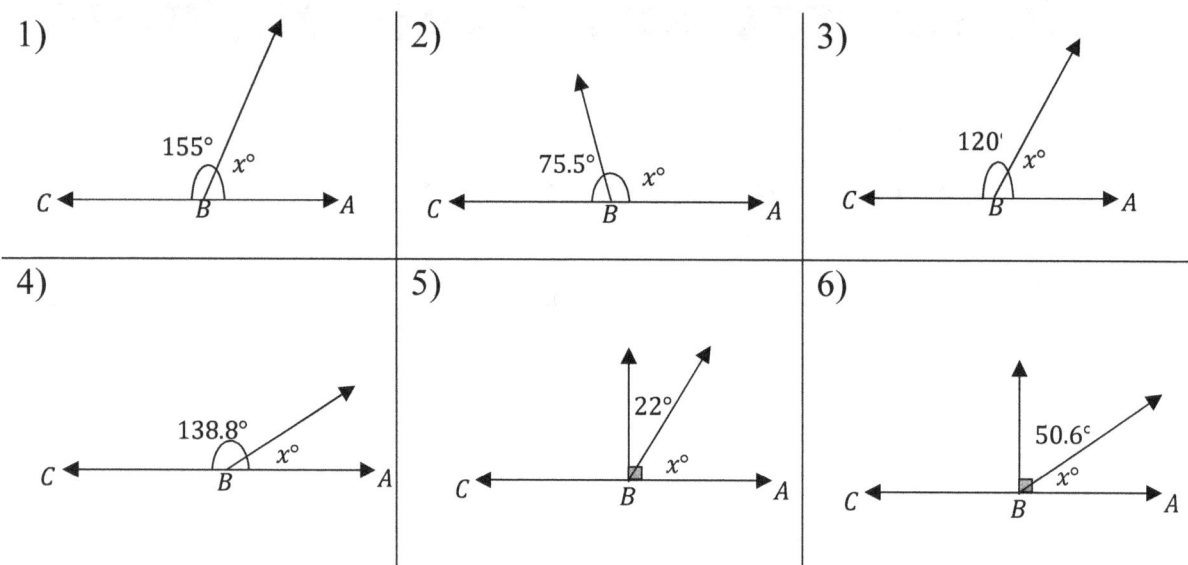

✏️ **Calculate.**

7) Two supplementary angles have equal measures. What is the measure of each angle? _____

8) The measure of an angle is nine seventh the measure of its supplement. What is the measure of the angle? _____

9) Two angles are complementary and the measure of one angle is 24 less than the other. What is the measure of the bigger angle? _____

10) Two angles are complementary. The measure of one angle is one fifth the measure of the other. What is the measure of the smaller angle? _____

11) Two supplementary angles are given. The measure of one angle is 80° less than the measure of the other. What does the bigger angle measure? _____

Pythagorean Relationship

✎ **Do the following lengths form a right triangle?**

1)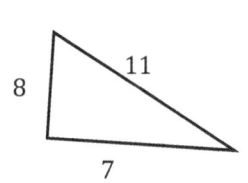

2)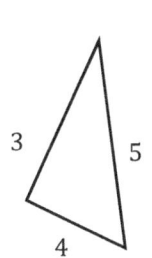

3)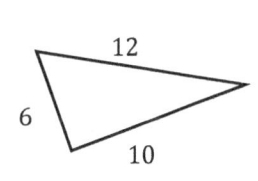

4)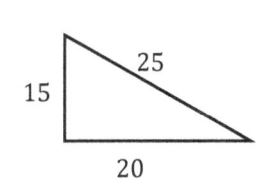

5)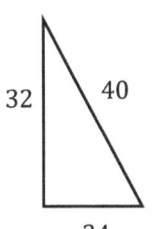

6)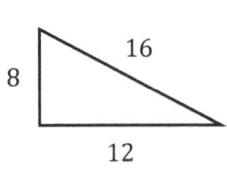

7)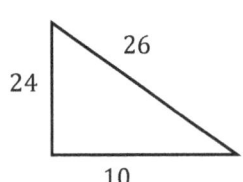

8)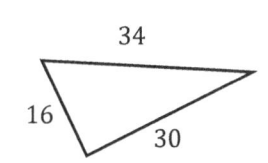

✎ **Find the missing side?**

9)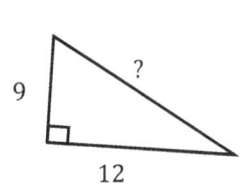

10)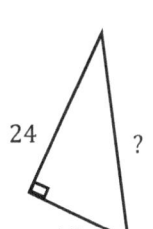

11)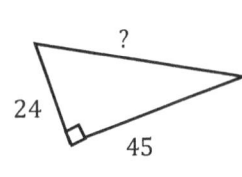

12)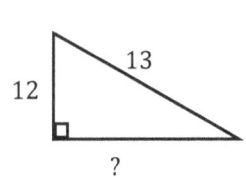

13)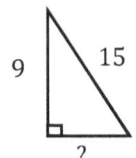

14)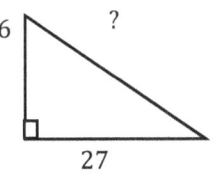

15)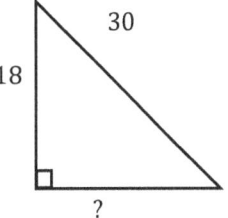

16)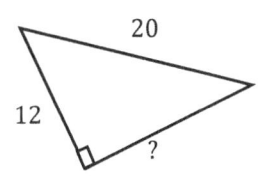

Triangles

✎ Find the measure of the unknown angle in each triangle.

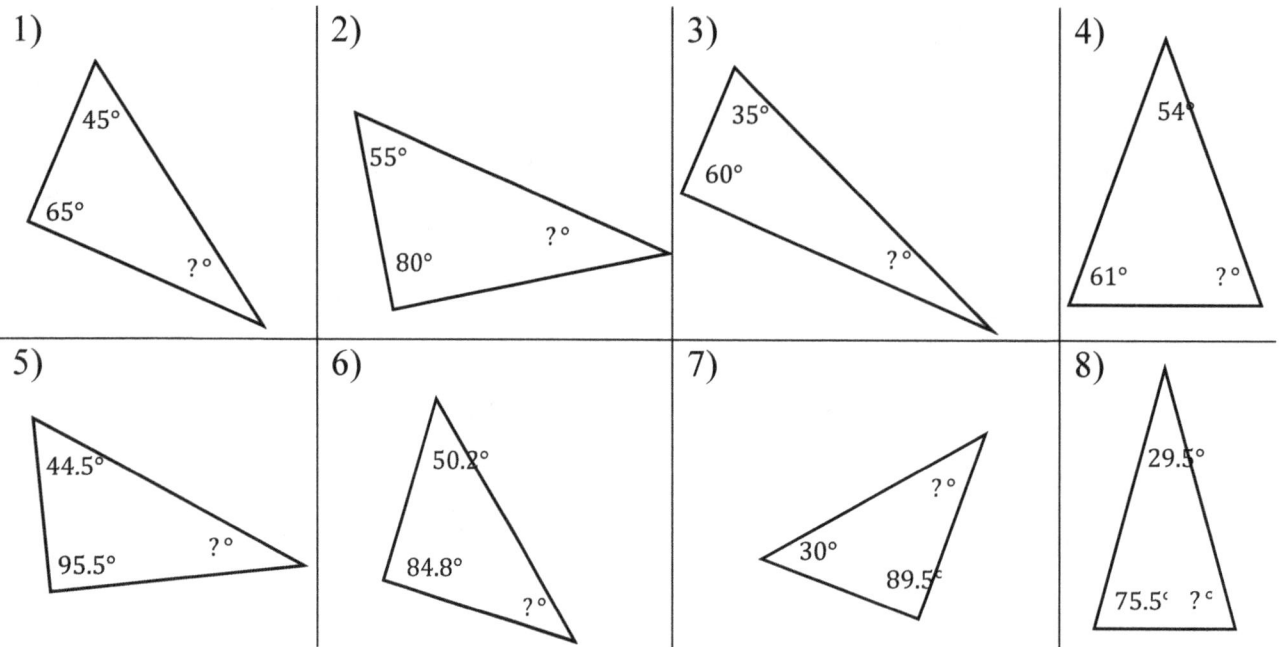

✎ Find area of each triangle.

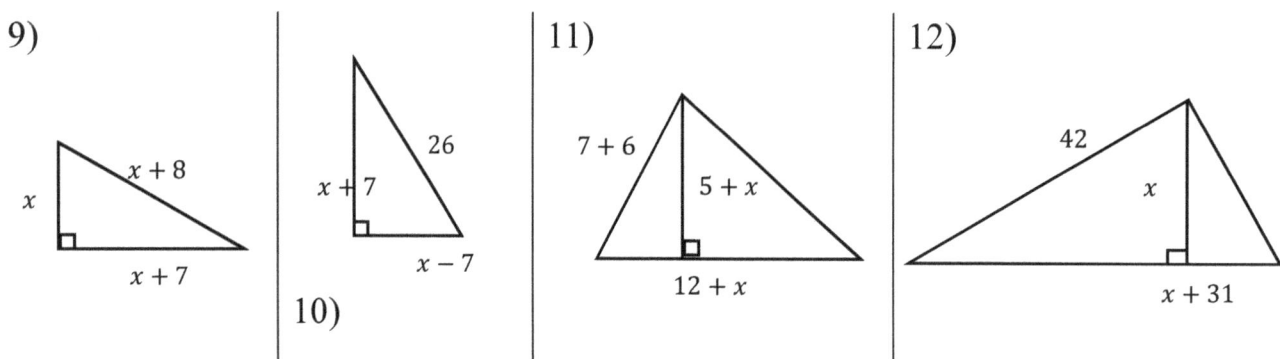

Polygons

👉 **Find the perimeter of each shape.**

1)

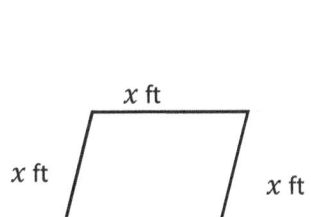

2)

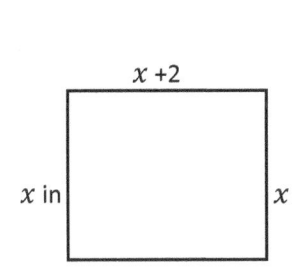

3)

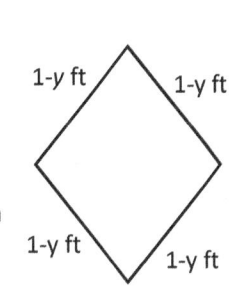

4) Square

(x + 1) cm

5) Regular hexagon

(2x) m

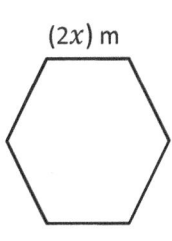

6)

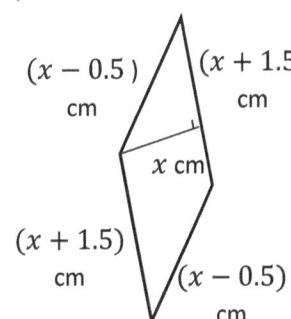

7) Parallelogram

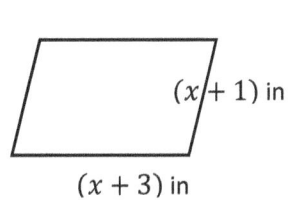

(x + 1) in
(x + 3) in

8) Square

(x + 2) m

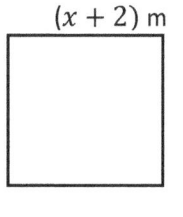

👉 **Find the area of each shape.**

9) Parallelogram

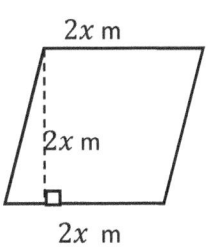

10) Rectangle

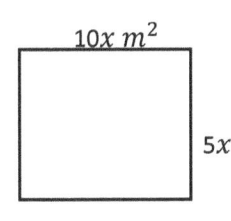

10x m^2
5x

11) Rectangle

(2+x) km
(2-x) km

12) Square

0.6x m

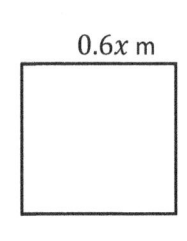

Trapezoids

✏️ **Find the area of each trapezoid.**

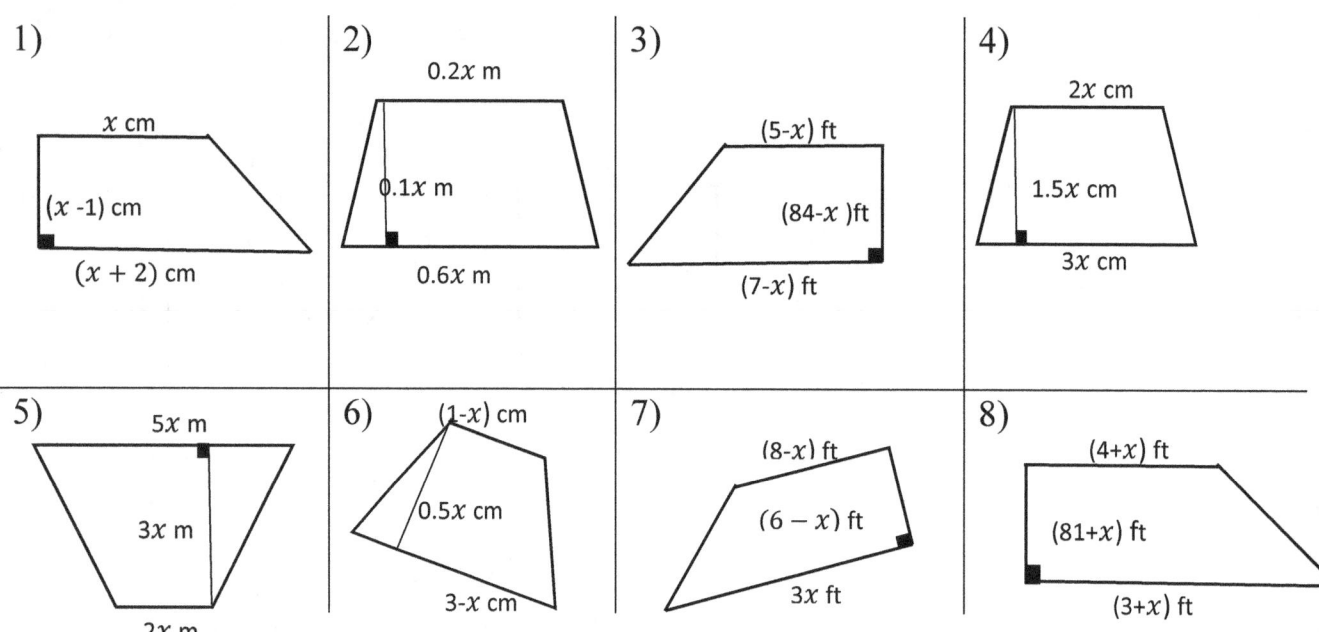

✏️ **Calculate.**

1) A trapezoid has an area of 40 cm² and its height is 8 cm and one base is 6 cm. What is the other base length? _____

2) If a trapezoid has an area of 85 ft² and the lengths of the bases are 9 ft and 8 ft, find the height. _____

3) If a trapezoid has an area of 150 m² and its height is 15 m and one base is 9 m, find the other base length. _____

4) The area of a trapezoid is 196 ft² and its height is 14 ft. If one base of the trapezoid is 12 ft, what is the other base length? _____

Circles

Find the area of each circle. ($\pi = 3.14$)

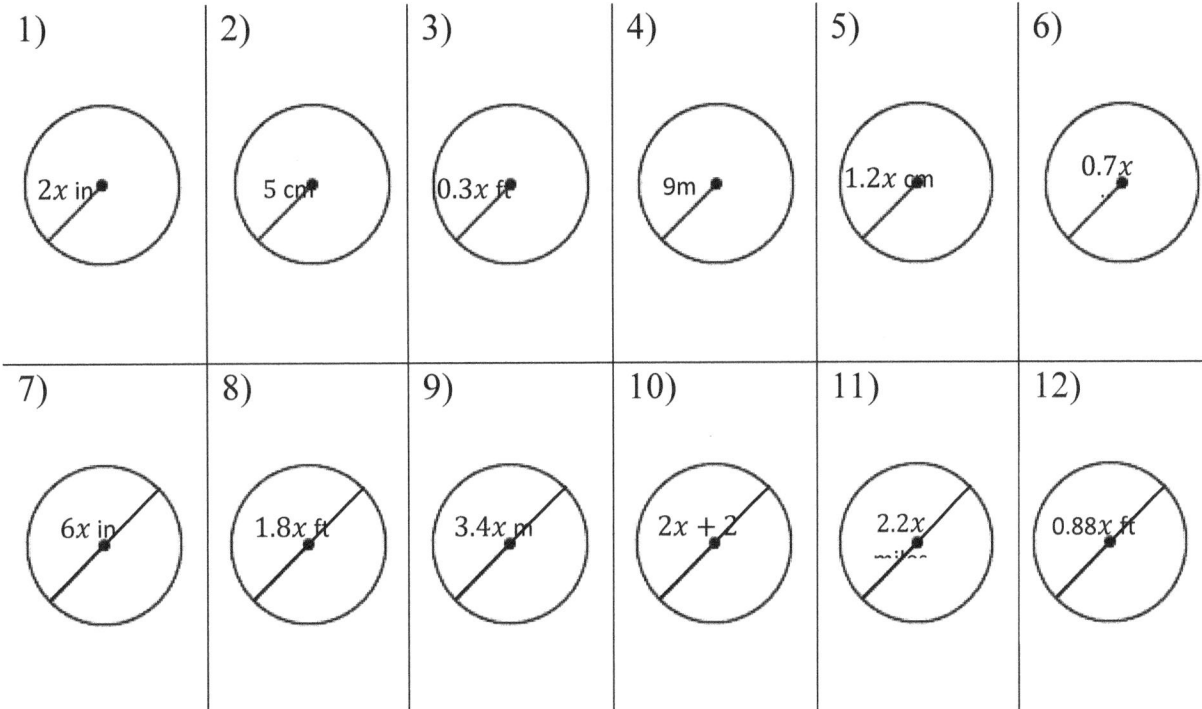

1) $2x$ in
2) 5 cm
3) $0.3x$ ft
4) 9 m
5) $1.2x$ cm
6) $0.7x$
7) $6x$ in
8) $1.8x$ ft
9) $3.4x$ m
10) $2x + 2$
11) $2.2x$ miles
12) $0.88x$ ft

Complete the table below. ($\pi = 3.14$)

Circle No.	Radius	Diameter	Circumference	Area
1	1.4 inches	2.8 inches	8.792 inches	6.154 square inches
2		4.6 meters		
3				$2.01x^2$ square ft
4			36.42 miles	
5		$6.2x$ kilometers		
6	$5x$ centimeters			
7		$2x$ feet		
8				1.54 square meters
9			$5.7x$ inches	
10	$(1-x)$ feet			

Cubes

✏️ **Find the volume of each cube.**

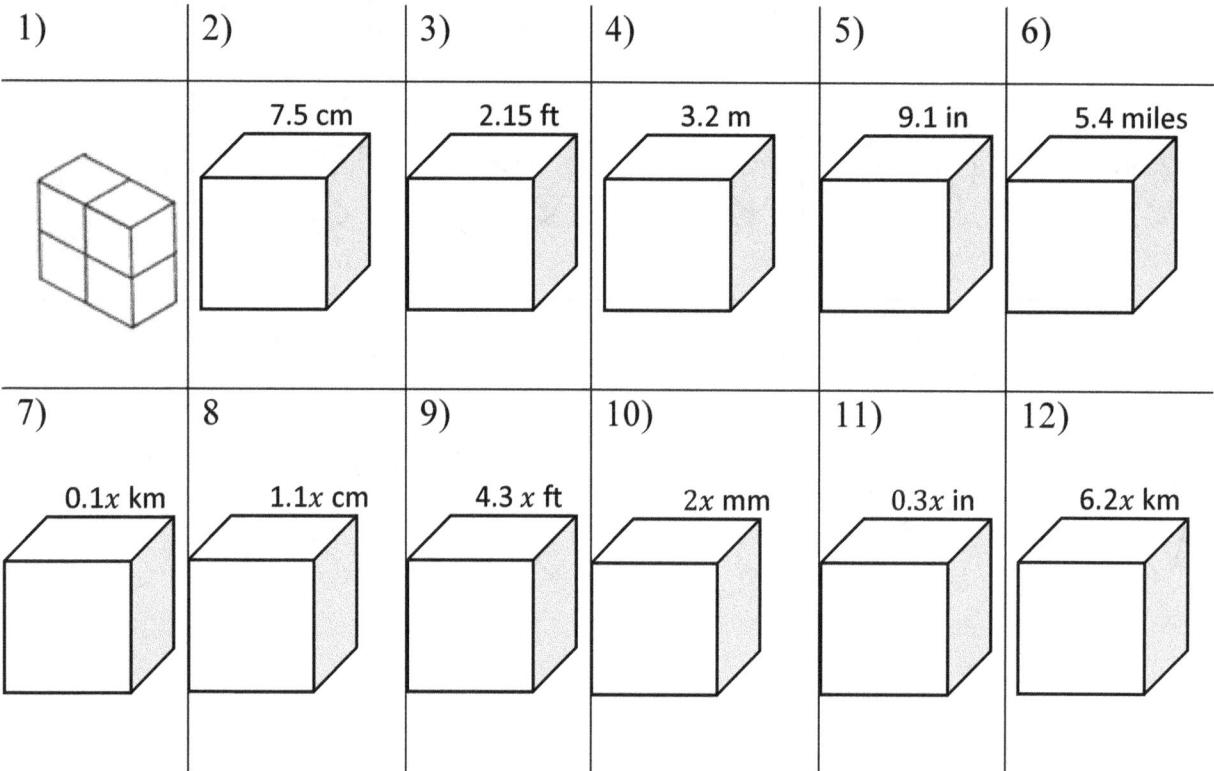

1)
2) 7.5 cm
3) 2.15 ft
4) 3.2 m
5) 9.1 in
6) 5.4 miles

7) $0.1x$ km
8) $1.1x$ cm
9) $4.3x$ ft
10) $2x$ mm
11) $0.3x$ in
12) $6.2x$ km

✏️ **Find the surface area of each cube.**

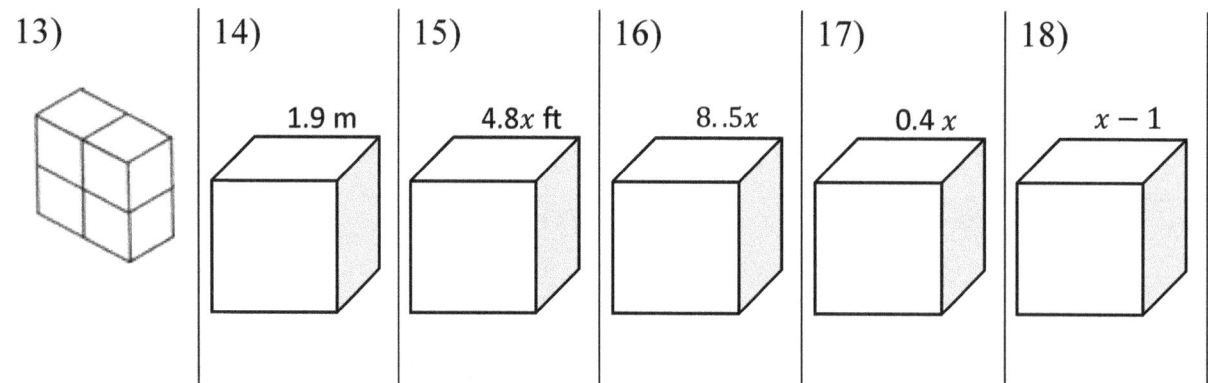

13)
14) 1.9 m
15) $4.8x$ ft
16) $8..5x$
17) $0.4x$
18) $x-1$

Rectangular Prism

✏️ **Find the volume of each Rectangular Prism.**

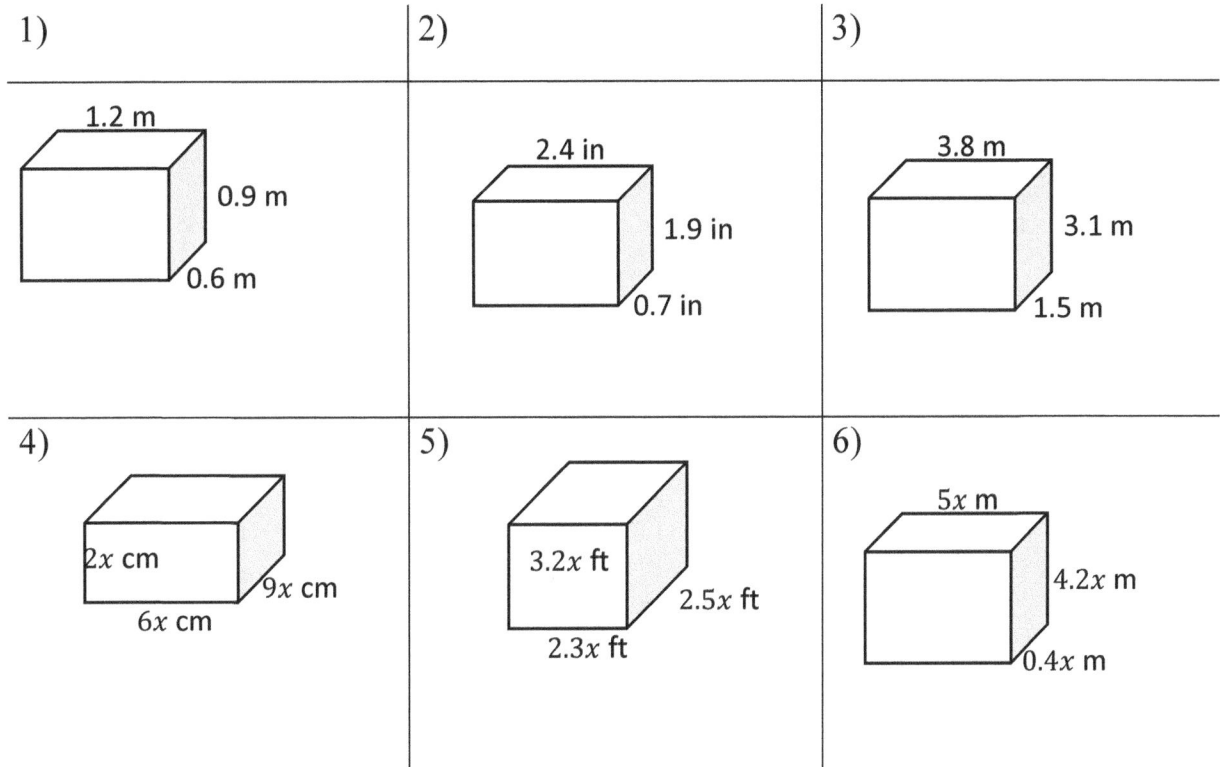

✏️ **Find the surface area of each Rectangular Prism.**

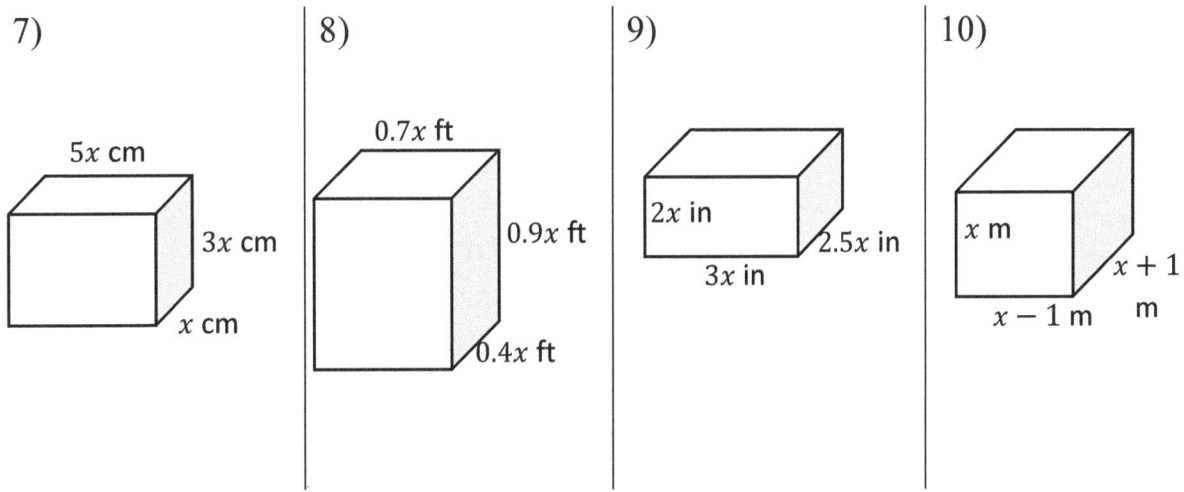

Cylinder

✏️ **Find the volume of each Cylinder. Round your answer to the nearest tenth.** ($\pi = 3.14$)

1)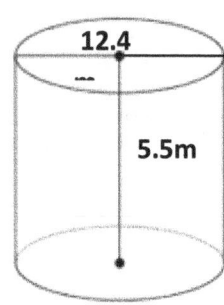

2)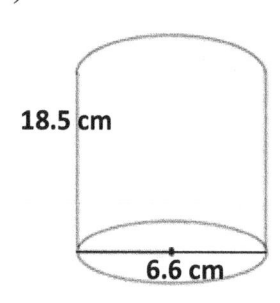

3)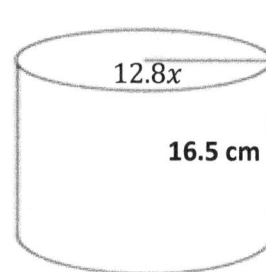

4)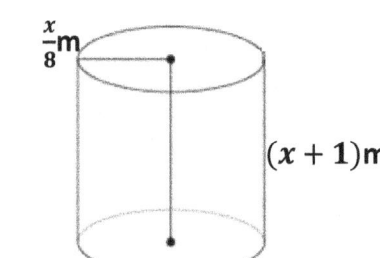

5)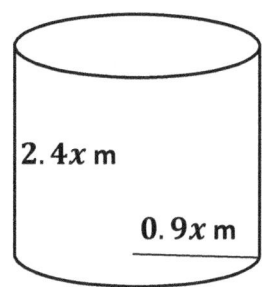

6)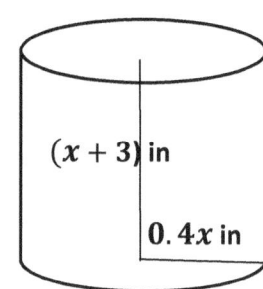

✏️ **Find the surface area of each Cylinder.** ($\pi = 3.14$)

7)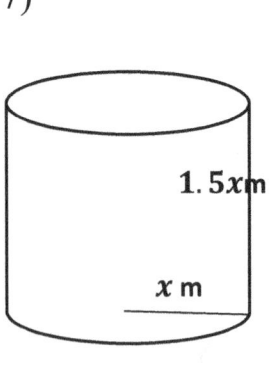

8)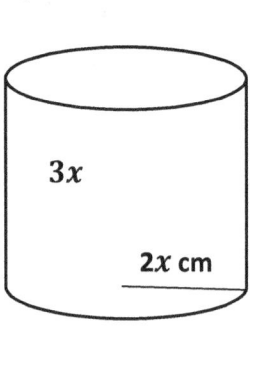

9)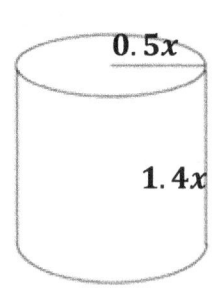

10)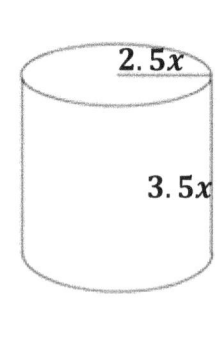

Pyramids and Cone

🖎 **Find the volume of each Pyramid and Cone.** ($\pi = 3.14$)

1)
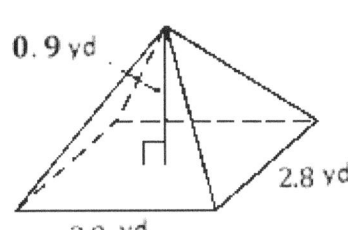

2)
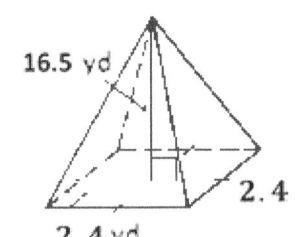

3)
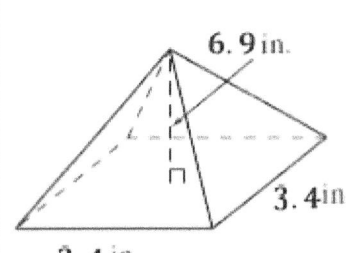

4)
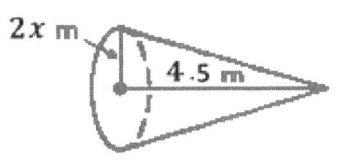

5)
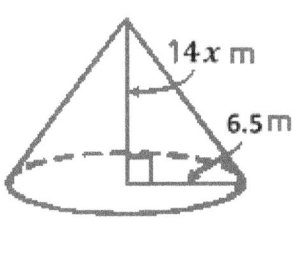

6)
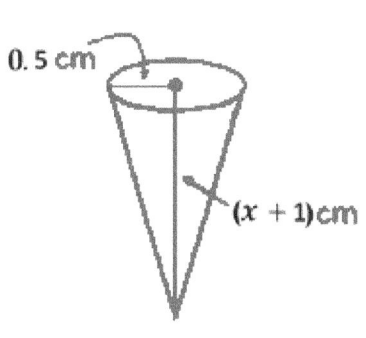

🖎 **Find the surface area of each Pyramid and Cone.** ($\pi = 3.14$)

7)
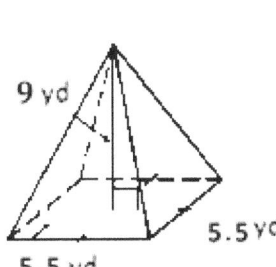

8)
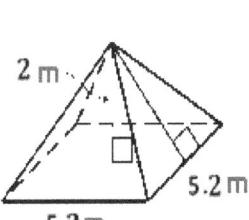

9)
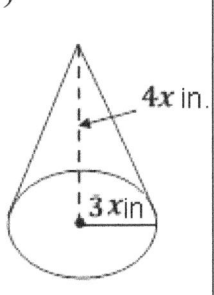

10)
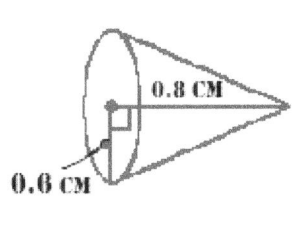

Answers of Worksheets – Chapter 13

Angles

1) 25°
2) 104.5°
3) 60°
4) 41.2°
5) 68°
6) 39.4°
7) 90°
8) 101.25°
9) 57°
10) 15°
11) 130°

Pythagorean Relationship

1) No
2) Yes
3) No
4) Yes
5) Yes
6) No
7) Yes
8) Yes
9) 15
10) 30
11) 51
12) 5
13) 12
14) 45
15) 24
16) 16

Triangles

1) 70°
2) 45°
3) 85°
4) 65°
5) 40°
6) 45°
7) 60.5°
8) 75°
9) $(\frac{x^2+7x}{2})$ square unites
10) $(\frac{x^2-49}{2})$ square unites
11) $(\frac{x^2+17x+60}{2})$ square unites
12) $(\frac{x^2+31x}{2})$ square unites

Polygons

1) $(4x)\ ft$
2) $(4x + 4)\ in$
3) $(4 - 4y)\ ft$
4) $(4x + 4)\ cm$
5) $(12x)\ m$
6) $(4x + 2)\ cm$
7) $(4x + 8)\ in$
8) $(4x + 8)\ m$
9) $(4x^2)m^2$
10) $(50x^2)m^2$
11) $(4 - x^2)\ km^2$
12) $(0.36x^2)\ m^2$

Trapezoids

1) $(x^2 - 1)\ cm^2$
2) $(0.04x^2)\ m^2$
3) $(x^2 - 10x + 24)\ ft^2$
4) $(3.75x^2)\ cm^2$
5) $(10.5x^2)m^2$
6) $(x - 0.5x^2)cm^2$
7) $(-x^2 + 2x + 24)\ ft^2$
8) $(\frac{2x^2+9x+7}{2})ft^2$

Calculate

1) 4 cm
2) 10 ft
3) 11 m
4) 16 ft

Circles

1) $(12.56x^2)\ in^2$
2) $78.5\ cm^2$
3) $(0.283x^2)\ ft^2$
4) $254.34 m^2$
5) $(4.522x^2)cm^2$
6) $(1.54x^2)\ miles^2$
7) $(28.56x^2)\ in^2$
8) $(2.543x^2)ft^2$
9) $(9.075x^2)\ m^2$

CLEP College Math Workbook

10) $(3.14x^2 + 6.28x + 3.14)\ cm^2$ 11) $(3.8x^2)\ miles^2$ 12) $(0.608x^2)\ ft^2$

Circle No.	Radius	Diameter	Circumference	Area
1	1.4 inches	2.8 inches	8.792 inches	6.154 square inches
2	2.3 meters	4.6 meters	14.44 meters	16.61 meters
3	$0.8x$ square ft	$1.6x$ square ft	$5.024x$ square ft	$2.01x^2$ square ft
4	5.8 miles	11.6 miles	36.42 miles	105.63 miles
5	$3.1x$ kilometers	$6.2x$ kilometers	$19.47x$ kilometers	$30.175x^2$ kilometers
6	$5x$ centimeters	$10x$ centimeters	$31.4x$ centimeters	$78.5x^2$ centimeters
7	x feet	$2x$ feet	$6.28x$ feet	$3.14x^2$ feet
8	0.7 square meters	1.4 square meters	4.396 square meters	1.54 square meters
9	$2.5x$ inches	$5x$ inches	$15.7x$ inches	$19.625x^2$ inches
10	$(1-x)$ feet	$2-2x$ feet	$6.28 - 6.28x)$ feet	$3.14x^2 - 6.28x + 3.14)$ feet

Cubes

1) 4
2) $421.88\ cm^3$
3) $9.94\ ft^3$
4) $32.77\ m^3$
5) $753.57\ in^3$
6) $157.46\ miles^3$
7) $(0.001x^3)\ km^3$
8) $(1.33x^3)\ cm^3$
9) $(79.51x^3)\ ft^3$
10) $(8x^3)\ mm^3$
11) $(0.027x^3)\ in^3$
12) $(238.33x^3)\ km^3$
13) 12
14) $21.66\ m^2$
15) $(138.24x^2)\ ft^2$
16) $(433.5x^2)\ mm^2$
17) $(0.96x^2)\ km^2$
18) $6x^2 - 12x + 6\ cm^2$

Rectangular Prism

1) $0.65\ m^3$
2) $3.19\ in^3$
3) $17.67\ m^3$
4) $(108x^3)\ cm^3$
5) $(18.4x^3)\ ft^3$
6) $(8.4x^3)\ m^3$
7) $(46x^2)\ cm^2$
8) $(2.54x^2)\ ft^2$
9) $(37x^2)\ in^2$
10) $(6x^2 - 2)\ m^2$

Cylinder

1) $663.86\ m^3$
2) $632.6\ cm^3$
3) $(8,488.55x^2)\ cm^3$
4) $(0.05x^3 + 0.05x^2)\ m^3$
5) $(6.104x^3)\ m^3$
6) $(0.5\ x^3 + 1.51x^2)\ in^3$
7) $(15.7x^2)\ m^2$
8) $(62.8x^2)\ cm^2$
9) $(5.97x^2)\ cm^2$
10) $(94.2x^2)\ m^2$

Pyramids and Cone

1) $2.35\ yd^3$
2) $31.68\ yd^3$
3) $26.59\ in^3$
4) $(18.84x^2)\ m^3$
5) $(619.10x)\ m^3$
6) $(0.262x + 0.262)\ cm^3$
7) $133.77\ yd^2$
8) $61.15\ m^2$
9) $(75.36x^2)\ in^2$
10) $(3.01)\ cm^2$

Chapter 14:
Statistics and Probability

Topics that you will practice in this chapter:

- ✓ Mean and Median
- ✓ Mode and Range
- ✓ Probability Problems
- ✓ Factorials
- ✓ Combinations and Permutation

Mathematics is no more computation than typing is literature.

– John Allen Paulos

Mean and Median

✏️ **Find Mean and Median of the Given Data.**

1) 8, 9, 19, 3, 4

2) 11, 7, 35, 10, 17, 32, 24

3) 38, 9, 15, 17, 13

4) 50, 19, 2, 18, 6, 7

5) 25, 27, 13, 16, 6, 13, 54

6) 24, 364, 42, 57, 6, 68

7) 89, 98, 65, 45, 3, 4, 30, 42

8) 34, 15, 15, 17, 22, 29, 15

9) 2, 5, 10, 45, 8, 13, 35, 6

10) 20, 22, 18, 7, 2, 17, 44, 53

11) 33, 52, 81, 9, 45, 31

12) 19, 74, 51, 8, 12, 15, 9, 14

✏️ **Calculate.**

13) In a javelin throw competition, five athletics score 45, 33, 53, 46 and 19 meters. What are their Mean and Median? _____

14) Eva went to shop and bought 5 apples, 9 peaches, 4 bananas, 7 pineapples and 8 melons. What are the Mean and Median of her purchase? _____

15) Bob has 19 black pen, 15 red pen, 27 green pens, 21 blue pens and one boxes of yellow pens. If the Mean and Median are 19 respectively, what is the number of yellow pens in box? _____

Mode and Range

✎ **Find Mode and Rage of the Given Data.**

1) 7, 4, 18, 9, 9, 3
   Mode: _____   Range: _____

2) 8, 8, 15, 14, 8, 5, 6, 18
   Mode: _____   Range: _____

3) 4, 4, 4, 15, 19, 24, 31, 5, 4
   Mode: _____   Range: _____

4) 10, 10, 9, 17, 14, 8, 20, 4
   Mode: _____   Range: _____

5) 5, 11, 3, 4, 3, 3
   Mode: _____   Range: _____

6) 13, 7, 7, 7, 7, 4, 12, 25, 8, 3
   Mode: _____   Range: _____

7) 1, 7, 9, 9, 24, 24, 24, 20, 34, 35
   Mode: _____   Range: _____

8) 9, 4, 7, 13, 13, 13, 9, 8, 15
   Mode: _____   Range: _____

9) 8, 8, 8, 5, 8, 7, 17, 16, 3, 9
   Mode: _____   Range: _____

10) 34, 34, 32, 14, 6, 14, 9, 14
    Mode: _____   Range: _____

11) 8, 8, 6, 8, 18, 10, 16, 15
    Mode: _____   Range: _____

12) 12, 12, 7, 11, 14, 12, 33, 5
    Mode: _____   Range: _____

✎ **Calculate.**

13) A stationery sold 21 pencils, 42 red pens, 25 blue pens, 26 notebooks, 21 erasers, 28 rulers and 27 color pencils. What are the Mode and Range for the stationery sells?

   Mode: _____   Range: _____

14) In an English test, eight students score 19, 10, 10, 17, 35, 35, 14 and 10. What are their Mode and Range? _____

15) What is the range of the first 6 odd numbers greater than 8?
   _____

Probability Problems

✏ **Calculate.**

1) A number is chosen at random from 1 to 20. Find the probability of selecting number 8 or smaller numbers. _____

2) Bag A contains 16 red marbles and 6 green marbles. Bag B contains 12 black marbles and 18 orange marbles. What is the probability of selecting a green marble at random from bag A? What is the probability of selecting a black marble at random from Bag B? _____

3) A number is chosen at random from 1 to 25. What is the probability of selecting multiples of 5? _____

4) A card is chosen from a well-shuffled deck of 52 cards. What is the probability that the card will be a queen? _____

5) A number is chosen at random from 1 to 15. What is the probability of selecting a multiple of 4? _____

A spinner, numbered 1–8, is spun once. What is the probability of spinning …?

6) an Odd number? _____   7) a multiple of 2? _____

8) a multiple of 5? _____   9) number 10? _____

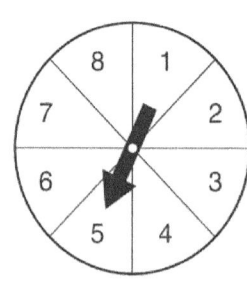

Factorials

✒ **Determine the value for each expression.**

1) $6! + 1! =$

2) $5! + 2! =$

3) $(4!)^2 =$

4) $6! - 3! =$

5) $8! - 4! + 3 =$

6) $3! \times 4 - 12 =$

7) $(3! + 1!)^2 =$

8) $(5! - 4!)^2 =$

9) $(3! \, 0!)^2 - 2 =$

10) $\dfrac{8!}{6!} =$

11) $\dfrac{3!}{2!} =$

12) $\dfrac{6!}{5!} =$

13) $\dfrac{21!}{19!} =$

14) $\dfrac{(n-1)!}{(n-3)!} =$

15) $\dfrac{(n+2)!}{(n+1)!} =$

16) $\dfrac{(4+2!)^3}{2!} =$

17) $\dfrac{4n!}{2n!} =$

18) $\dfrac{31!}{29!2!} =$

19) $\dfrac{13!}{9!3!} =$

20) $\dfrac{6 \times 280!}{3(4 \times 70)!} =$

21) $\dfrac{30!}{31!2!} =$

22) $\dfrac{7!7!}{8!5!} =$

23) $\dfrac{12!11!}{9!10!} =$

24) $\dfrac{(2 \times 5)!}{1!9!} =$

25) $\dfrac{2!(6n-1)!}{(6n)!} =$

26) $\dfrac{n(4n+4)!}{(4n+5)!} =$

27) $\dfrac{(n+1)!(n)}{(n+2)!} =$

WWW.MathNotion.Com

Combinations and Permutations

✎ **Calculate the value of each.**

1) 6! = ____

2) 2! × 5! = ____

3) 4! = ____

4) 3! + 5! = ____

5) 7! = ____

6) 9! = ____

7) 3! + 3! = ____

8) 5! − 2! = ____

✎ **Find the answer for each word problems.**

9) Susan is baking cookies. She uses sugar, Vanilla and eggs. How many different orders of ingredients can she try? _____

10) Albert is planning for his vacation. He wants to go to museum, watch a movie, go to the beach, play volleyball and play football. How many ways of ordering are there for him? _____

11) How many 6-digit numbers can be named using the digits 1, 6, 8, 9, and 10 without repetition? _____

12) In how many ways can 4 boys be arranged in a straight line? _____

13) In how many ways can 8 athletes be arranged in a straight line? _____

14) A professor is going to arrange her 5 students in a straight line. In how many ways can she do this? _____

15) How many code symbols can be formed with the letters for the word FRIEND? _____

16) In how many ways a team of 7 basketball players can to choose a captain and co-captain? _____

Answers of Worksheets – Chapter 14

Mean and Median

1) Mean: 8.6, Median: 8
2) Mean: 19.43, Median: 17
3) Mean: 18.4, Median: 15
4) Mean: 17, Median: 12.5
5) Mean: 22, Median: 16
6) Mean: 93.5, Median: 49.5
7) Mean: 47, Median: 43.5
8) Mean: 21, Median: 17
9) Mean: 15.5, Median: 9
10) Mean: 22.88, Median: 19
11) Mean: 41.83, Median: 39
12) Mean: 25.25, Median: 14.5
13) Mean: 39.2, Median: 45
14) Mean: 6.6, Median: 7
15) 13

Mode and Range

1) Mode: 9, Range: 15
2) Mode: 8, Range: 13
3) Mode: 4, Range: 27
4) Mode: 10, Range: 16
5) Mode: 3, Range: 8
6) Mode: 7, Range: 22
7) Mode: 24, Range: 34
8) Mode: 13, Range: 11
9) Mode: 8, Range: 14
10) Mode: 14, Range: 28
11) Mode: 8, Range: 12
12) Mode: 12, Range: 28
13) Mode: 21, Range: 21
14) Mode: 10, Range: 25
15) 10

Probability Problems

1) $\frac{2}{5}$
2) $\frac{3}{11}, \frac{2}{5}$
3) $\frac{1}{5}$
4) $\frac{1}{13}$
5) $\frac{1}{5}$
6) $\frac{1}{2}$
7) $\frac{1}{2}$
8) $\frac{1}{8}$
9) 0

Factorials

1) 721
2) 122
3) 576
4) 714
5) 40,299
6) 12
7) 49
8) 9,216
9) 34
10) 56
11) 3
12) 6
13) 420
14) $(n-1)(n-2)$
15) $n+2$
16) 108
17) 2
18) 465
19) 2,860
20) 2
21) $\frac{1}{62}$
22) 5.25
23) 14,520
24) 10
25) $\frac{1}{3n}$
26) $\frac{n}{4n+5}$
27) $\frac{n}{n+2}$

Combinations and Permutations

1) 720
2) 240
3) 24
4) 126
5) 5,040
6) 362,880

7) 12
8) 118
9) 6
10) 120
11) 720
12) 24
13) 40,320
14) 120
15) 720
16) 42

CLEP College Mathematics Test Review

College-Level Examination Program (CLEP) is a series of 33 standardized tests that measures your knowledge of certain subjects. You can earn college credit at thousands of colleges and universities by earning a satisfactory score on a computer based CLEP exam.

The CLEP College Mathematics measures your knowledge of math topics generally taught in a college course for non-mathematics majors. It contains approximately 60 multiple choice questions to be answered in 90 minutes. Some of these questions are pretest questions that will not be scored. These 60 questions cover: sets, logic, the real number system, functions and graphing, probability and statistics, and additional topics from algebra and geometry. A scientific calculator is available to students during the entire testing time.

The CLEP College mathematics exam score ranges from 20 to 80 converting to A, B, C, or D based on this score. The letter grade is applied to your college course equivalent.

In this section, there are two complete CLEP College Mathematics Tests. Take these tests to see what score you'll be able to receive on a real CLEP College Mathematics test.

Time to Test

Time to refine your skill with a practice examination

Take practice CLEP College Mathematics Tests to simulate the test day experience. After you've finished, score your tests using the answer keys.

Before You Start

- You'll need a pencil, a calculator and a timer to take the test.

- For each question, there are four possible answers. Choose which one is best.

- It's okay to guess. There is no penalty for wrong answers.

- Use the answer sheet provided to record your answers.

- After you've finished the test, review the answer key to see where you went wrong.

Good Luck!

CLEP College Mathematics Test Answer Sheets

Remove (or photocopy) these answer sheets and use them to complete the practice tests.

CLEP College Mathematics Practice Test

1	Ⓐ Ⓑ Ⓒ Ⓓ	21	Ⓐ Ⓑ Ⓒ Ⓓ	41	Ⓐ Ⓑ Ⓒ Ⓓ
2	Ⓐ Ⓑ Ⓒ Ⓓ	22	Ⓐ Ⓑ Ⓒ Ⓓ	42	Ⓐ Ⓑ Ⓒ Ⓓ
3	Ⓐ Ⓑ Ⓒ Ⓓ	23	Ⓐ Ⓑ Ⓒ Ⓓ	43	Ⓐ Ⓑ Ⓒ Ⓓ
4	Ⓐ Ⓑ Ⓒ Ⓓ	24	Ⓐ Ⓑ Ⓒ Ⓓ	44	Ⓐ Ⓑ Ⓒ Ⓓ
5	Ⓐ Ⓑ Ⓒ Ⓓ	25	Ⓐ Ⓑ Ⓒ Ⓓ	45	Ⓐ Ⓑ Ⓒ Ⓓ
6	Ⓐ Ⓑ Ⓒ Ⓓ	26	Ⓐ Ⓑ Ⓒ Ⓓ	46	Ⓐ Ⓑ Ⓒ Ⓓ
7	Ⓐ Ⓑ Ⓒ Ⓓ	27	Ⓐ Ⓑ Ⓒ Ⓓ	47	Ⓐ Ⓑ Ⓒ Ⓓ
8	Ⓐ Ⓑ Ⓒ Ⓓ	28	Ⓐ Ⓑ Ⓒ Ⓓ	48	Ⓐ Ⓑ Ⓒ Ⓓ
9	Ⓐ Ⓑ Ⓒ Ⓓ	29	Ⓐ Ⓑ Ⓒ Ⓓ	49	Ⓐ Ⓑ Ⓒ Ⓓ
10	Ⓐ Ⓑ Ⓒ Ⓓ	30	Ⓐ Ⓑ Ⓒ Ⓓ	50	Ⓐ Ⓑ Ⓒ Ⓓ
11	Ⓐ Ⓑ Ⓒ Ⓓ	31	Ⓐ Ⓑ Ⓒ Ⓓ	51	Ⓐ Ⓑ Ⓒ Ⓓ
12	Ⓐ Ⓑ Ⓒ Ⓓ	32	Ⓐ Ⓑ Ⓒ Ⓓ	52	Ⓐ Ⓑ Ⓒ Ⓓ
13	Ⓐ Ⓑ Ⓒ Ⓓ	33	Ⓐ Ⓑ Ⓒ Ⓓ	53	Ⓐ Ⓑ Ⓒ Ⓓ
14	Ⓐ Ⓑ Ⓒ Ⓓ	34	Ⓐ Ⓑ Ⓒ Ⓓ	54	Ⓐ Ⓑ Ⓒ Ⓓ
15	Ⓐ Ⓑ Ⓒ Ⓓ	35	Ⓐ Ⓑ Ⓒ Ⓓ	55	Ⓐ Ⓑ Ⓒ Ⓓ
16	Ⓐ Ⓑ Ⓒ Ⓓ	36	Ⓐ Ⓑ Ⓒ Ⓓ	56	Ⓐ Ⓑ Ⓒ Ⓓ
17	Ⓐ Ⓑ Ⓒ Ⓓ	37	Ⓐ Ⓑ Ⓒ Ⓓ	57	Ⓐ Ⓑ Ⓒ Ⓓ
18	Ⓐ Ⓑ Ⓒ Ⓓ	38	Ⓐ Ⓑ Ⓒ Ⓓ	58	Ⓐ Ⓑ Ⓒ Ⓓ
19	Ⓐ Ⓑ Ⓒ Ⓓ	39	Ⓐ Ⓑ Ⓒ Ⓓ	59	Ⓐ Ⓑ Ⓒ Ⓓ
20	Ⓐ Ⓑ Ⓒ Ⓓ	40	Ⓐ Ⓑ Ⓒ Ⓓ	60	Ⓐ Ⓑ Ⓒ Ⓓ

WWW.MathNotion.Com

CLEP College Mathematics

Practice Test 1

- ❖ 60 Questions.
- ❖ Total time for this test: 90 Minutes.
- ❖ You may use a scientific calculator on this test.

Administered *Month Year*

CLEP College Math Workbook

1) If a, b and c are positive integers and $4a = 5b = 3c$, then the value of $2a + 5b + 6c$ is how many times the value of a?

   A. 14

   B. 10

   C. 15.5

   D. 16

2) If $f(x^2) = 6x + 4$, for all positive value of x, what is the value of $f(144)$?

   A. -76

   B. 76

   C. 26

   D. -26

3) If a and b are solutions of the following equation, which of the following is the ratio $\frac{a}{b}$? $(a > b)$

$$3x^2 - 22x + 16 = -7x + 34$$

   A. $\frac{1}{6}$

   B. -6

   C. $-\frac{1}{6}$

   D. 6

4) If $x \neq -7$ and $x \neq 6$, which of the following is equivalent to $\frac{1}{\frac{1}{x-6}+\frac{1}{x+7}}$?

   A. $\frac{(x-6)(x+7)}{(x-6)+(x+7)}$

   B. $\frac{(x+7)+(x-6)}{(x+7)(x-6)}$

   C. $\frac{(x+7)(x-6)}{(x+7)-(x+6)}$

   D. $\frac{(x+7)+(x-6)}{(x+7)-(x-6)}$

5) A line in the xy-plane passes through origin and has a slope of $\frac{1}{2}$. Which of the following points lies on the line?

   A. $(1,4)$

   B. $(4,1)$

   C. $(4, 2)$

   D. $(2,4)$

6) Which of the following is the solution of the following inequality?

$$4x + 4..5 > 12x - 9.5 - 4.5x$$

A. $x < 4$

B. $x > 4$

C. $x \leq 5$

D. $x \geq 5$

Gender	Under 55	55 or older	total
Male	9	11	20
Female	13	7	20
Total	22	18	40

7) The table above shows the distribution of age and gender for 40 employees in a company. If one employee is selected at random, what is the probability that the employee selected be either a female under age 55 or a male age 55 or older?

A. $\frac{11}{13}$

B. $\frac{24}{40}$

C. $\frac{11}{40}$

D. $\frac{13}{40}$

8) If a parabola with equation $y = ax^2 + 4x + 22$, where a is constant, passes through point (3, 7), what is the value of a^2?

A. -3

B. 3

C. -9

D. 9

9) John works for an electric company. He receives a monthly salary of $3,400 plus 8% of all his monthly sales as bonus. If x is the number of all John's sales per month, which of the following represents John's monthly revenue in dollars?

A. $0.08x$

B. $0.92x - 3,400$

C. $0.08x + 3,400$

D. $0.92x + 3,400$

10) What is the value of $f(3)$ for the following function f?

$$f(x) = x^3 - 9x$$

A. 2

B. 0

C. 3

D. 9

11) John buys a pepper plant that is 7 inches tall. With regular watering the plant grows 5 inches a year. Writing John's plant's height as a function of time, what does the y-intercept represent?

A. The y-intercept represents the rate of grows of the plant which is 7 in.

B. The y-intercept represents the starting height of 7 in.

C. The y-intercept represents the rate of growth of plant which is 5 in. per year

D. There is no y-intercept

12) What is the length of AB in the following figure if $AE = 3$, $CD = 2$ and $AC = 15$?

A. 45

B. 6

C. 10

D. 9

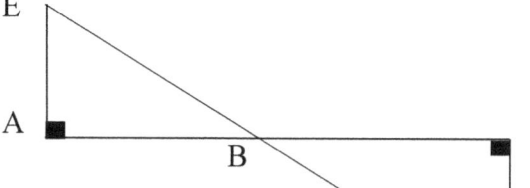

CLEP College Math Workbook

13) What is the solution of the following system of equations?

$$\begin{cases} \frac{-x}{6} + \frac{y}{3} = 2 \\ \frac{-6}{8}x + 2y = 12 \end{cases}$$

A. $x = 4, y = 6$

B. $x = 0\ y = 6$

C. $x = 2, y = 12$

D. $x = 6, y = 0$

14) If $x \neq 0$, what is the value of $\frac{(9(x^2)(y^3))^3}{(6xy^2)^3}$?

A. $\frac{9}{6}x^2y^3$

B. $\frac{27}{8}x^3y^5$

C. $(\frac{3}{2}xy)^3$

D. $\frac{27}{8}(xy)^2$

15) In the following equation, what is the value of $y - 5x$?

$$\frac{y}{8} = x - \frac{3}{8}x + 9$$

A. 72

B. 24

C. 36

D. 18

16) The length of a rectangle is 2 meters greater than 7 times its width. The perimeter of the rectangle is 36 meters. What is the area of the rectangle in meters?

A. 14

B. 32

C. 18

D. 12

CLEP College Math Workbook

17) If $y = nx + 5$, where n is a constant, and when $x = 8$, $y = 12$, what is the value of y when $x = 8$?

   A. 39

   B. 16

   C. 20

   D. 13

18) If $6 + 3x$ is 12 more than 18, what is the value of $9x$?

   A. 30

   B. 36

   C. 64

   D. 72

19) If a gas tank can hold 30 gallons, how many gallons does it contain when it is $\frac{4}{5}$ full?

   A. 48

   B. 96

   C. 24

   D. 36

20) In the xy-plane, the point $(3, 7)$ and $(2, 6)$ are online A. Which of the following equations of lines is parallel to line A?

   A. $y = 4x$

   B. $y = \frac{x}{4}$

   C. $y = x$

   D. $y = 5x$

21) A football team won exactly 60% of the games it played during last session. Which of the following could be the total number of games the team played last season?

   A. 42

   B. 40

   C. 46

   D. 29

CLEP College Math Workbook

22) The capacity of a red box is 40% bigger than the capacity of a blue box. If the red box can hold 35 equal sized books, how many of the same books can the blue box hold?

A. 10

B. 15

C. 20

D. 25

23) The sum of five different negative integers is -65. If the smallest of these integers is -15, what is the largest possible value of one of the other four integers?

A. -13

B. -12

C. -11

D. -24

24) If x is greater than 1 and less than 2, which of the following is true?

A. $x < \sqrt{x^2 + 1} < \sqrt{x^2} + 1$

B. $x < \sqrt{x^2} + 1 < \sqrt{x^2 + 1}$

C. $\sqrt{x^2 + 1} < x < \sqrt{x^2} + 1$

D. $\sqrt{x^2} + 1 < \sqrt{x^2 + 1} < x$

25) The ratio of boys and girls in a class is 3:7. If there are 90 students in the class, how many more boys should be enrolled to make the ratio 1:1?

A. 27

B. 9

C. 36

D. 7

26) If $f(x) = 4x + 2(x + 2) + 2$ then $f(3x) = ?$

A. $18x + 6$

B. $18x - 2$

C. $16x + 6$

D. $16x + 3$

CLEP College Math Workbook

Questions 27, 28 and 29 are based on the following data

Types of air pollutions in 10 cities of a country

Type of Pollution	Number of Cities									
A										
B										
C										
D										
E										
	1	2	3	4	5	6	7	8	9	10

27) If a is the mean (average) of the number of cities in each pollution type category, b is the mode, and c is the median of the number of cities in each pollution type category, then which of the following must be true?

A. $b < a < c$

B. $b < c < a$

C. $a = c$

D. $c < b = a$

28) How many cities should be added to type of pollutions C until the ratio of cities in type of pollution C to cities in type of pollution D will be 0.50?

A. 6

B. 2

C. 1

D. 3

29) What percent of cities are in the type of pollution B, C, and E respectively?

A. 50%, 60%, 30%

B. 1.30%, 1.20%, 1.50%

C. 1.80%, 1.60%, 1.50%

D. 30%, 20%, 50%

WWW.MathNotion.Com

30) In the following right triangle, if the sides AB and BC become tripe longer, what will be the ratio of the perimeter of the triangle to its area?

A. $\frac{1}{3}$

B. 4

C. $\frac{1}{4}$

D. 6

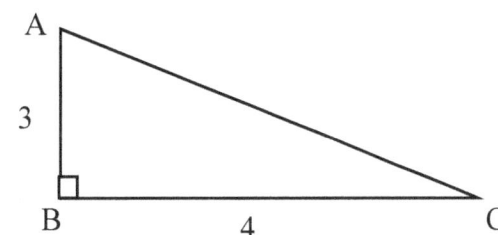

31) What is the ratio of the minimum value to the maximum value of the following function?

$$-1 \leq x \leq 2$$
$$f(x) = -5x + 2$$

A. $\frac{12}{7}$

B. $-\frac{8}{7}$

C. $-\frac{12}{7}$

D. $\frac{8}{7}$

32) If $\frac{a-b}{b} = \frac{5}{9}$, then which of the following must be true?

A. $\frac{a}{b} = \frac{13}{9}$

B. $\frac{a}{b} = \frac{14}{9}$

C. $\frac{a}{b} = \frac{12}{19}$

D. $\frac{a}{b} = \frac{14}{10}$

33) What is the value of x in the following equation?

$$\frac{x^2-16}{x+4} + 3(x+3) = 12$$

A. 4

B. 3

C. $\frac{7}{4}$

D. $\frac{7}{3}$

Questions 34 to 36 are based on the following data

The result of a research shows the number of men and women in four cities of a country.

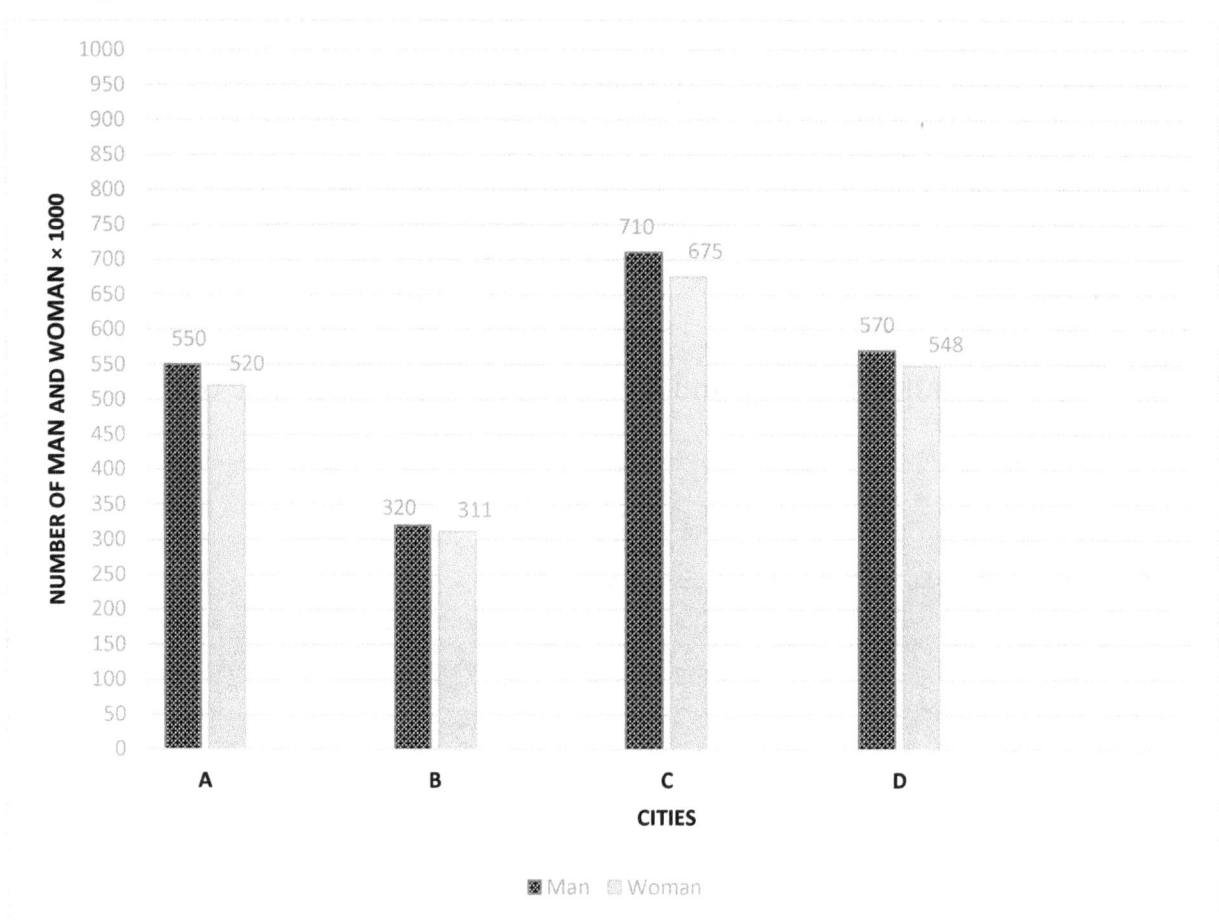

34) What's the ratio of percentage of men in city C to percentage of men in city B?

    A. 0.99   C. 0.99

    B. 0.11   D. 99.11

35) What's the minimum ratio of woman to man in the four cities?

    A. 0.96   C. 0.95

    B. 0.99   D. 0.94

36) How many women should be added to city D until the ratio of women to men will be 1.4?

   A. 128

   B. 228

   C. 248

   D. 148

37) In the rectangle below if $y > 4$ cm and the area of rectangle is 32 cm² and the perimeter of the rectangle is 24 cm, what is the value of x and y respectively?

   A. 8, 4

   B. 4, 12

   C. 4, 8

   D. 8, 12

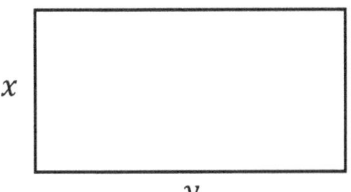

38) If a car has 80-liter petrol and after one hour driving the car use 5-liter petrol, how much petrol will remain after x-hours driving?

   A. $80x - 5$

   B. $80x + 5$

   C. $80 - 5x$

   D. $80 + 5x$

39) In the triangle below, if the measure of angle A is 35 degrees, then what is the value of y? (figure is NOT drawn to scale)

   A. 65

   B. 60

   C. 85

   D. 83

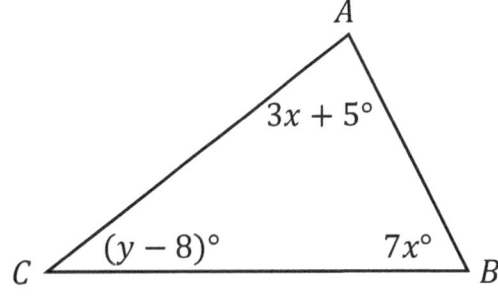

40) The following graph shows the mark of six students in mathematics. What is the mean (average) of the marks?

A. 12.4

B. 13

C. 13.5

D. 14

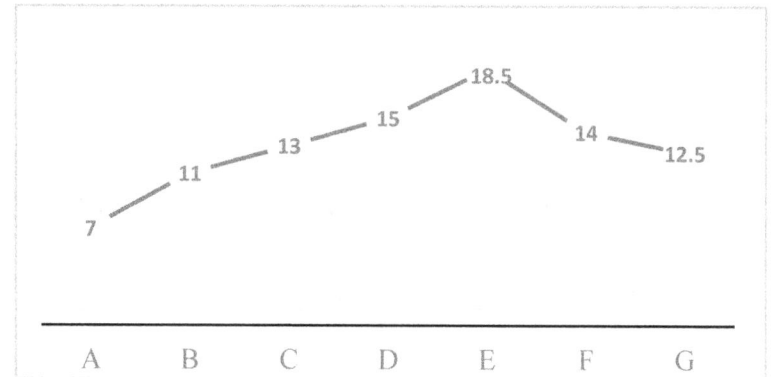

41) Which of the following values for x and y satisfy the following system of equations?

$$\begin{cases} x + 3y = 2 \\ -3x - 2y = 8 \end{cases}$$

A. $x = -2, y = 4$

B. $x = 2, y = 4$

C. $x = 4, y = -2$

D. $x = -4, y = 2$

42) Given the right triangle ABC bellow, sin (β) is equal to?

A. $\dfrac{b}{a}$

B. $\dfrac{b}{\sqrt{a^2+b^2}}$

C. $\dfrac{\sqrt{a^2+b^2}}{ab}$

D. $\dfrac{c}{\sqrt{a^2+b^2}}$

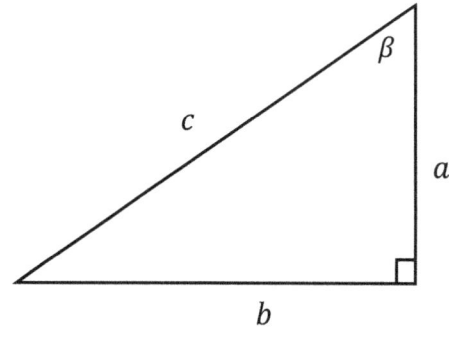

43) Solve the following inequality.

$$\left|\frac{x}{3} - x + 8\right| < 2$$

A. $-9 < x < 15$

B. $-15 < x < 9$

C. $9 < x < 15$

D. $-15 < x < -9$

44) If x is directly proportional to the square of y, and $y = 3$ when $x = 36$, then when $x = 256$ $y = ?$

A. $\frac{1}{8}$

B. 4

C. 8

D. 16

45) If $A = \{2, 8, 10, 11\}, B = \{1, 2, 3, 4, 8, 6\}$, and $C = \{8, 7, 9, 10, 12\}$, then which of the following set is $(A \cup B) \cap C$?

A. $\{1, 2, 3, 4, 8, 6, 11, 10\}$

B. $\{1, 2, 3, 4, 8, 6, 7, 11, 10, 12\}$

C. $\{8, 11, 12, 10\}$

D. $\{8, 10\}$

46) $f(x) = ax^2 + bx + c$ is a quadratic function where a, b and c are constant, the value of x of the point of intersection of this quadratic function and linear function $g(x) = 3x - 2$ is 2. The vertex of $f(x)$ is at $(-2, 0)$. What is the product of a, b and c?

A. $\frac{1}{4}$

B. $\frac{1}{2}$

C. $-\frac{1}{6}$

D. $-\frac{1}{8}$

CLEP College Math Workbook

47) A ladder leans against a wall forming a 60° angle between the ground and the ladder. If the bottom of the ladder is 24 feet away from the wall, how many feet is the ladder?

   A. 6 feet

   B. 12 feet

   C. 24 feet

   D. 48 feet

48) An angle is equal to one ninth of its supplement. What is the measure of that angle?

   A. 18

   B. 36

   C. 15

   D. 45

49) If $3x + 6y = \frac{-3y^2 + 18}{x}$, what is the value of $(x + y)^2$? $(x \neq 0)$

   A. 3

   B. 6

   C. 9

   D. 18

50) The volume of cube A is $\frac{1}{3}$ of its surface area. What is the length of an edge of cube A?

   A. 2

   B. 3

   C. 4

   D. 9

51) If $\sqrt{3m - 2} = m$, what is (are) the value(s) of m?

   A. 0

   B. −1, 2

   C. 1, 2

   D. −1, −2

52) If the function $g(x)$ has three distinct zeros, which of the following could represent the graph of $g(x)$?

A.
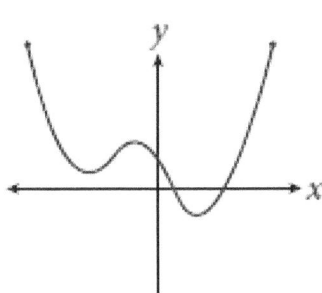

B.
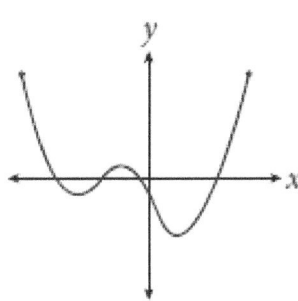

C.
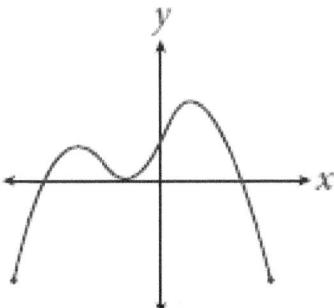

D.
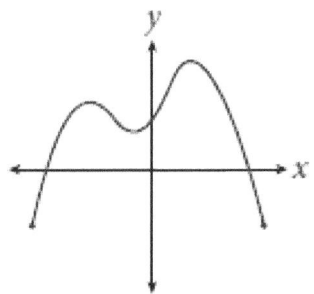

53) In the following equation when z is divided by 5, what is the effect on x?

$$x = \frac{8y + \frac{r}{r+1}}{\frac{6}{z}}$$

A. x is divided by $\frac{5}{6}$

B. x is divided by 5

C. x does not change

D. x is multiplied by $\frac{z}{r+1}$

54) A boat sails 40 miles south and then 30 miles east. How far is the boat from its start point?

A. 56 miles

B. 170 miles

C. 80 miles

D. 90 miles

55) For what real value of x the equation below is true?

$$x^3 - 5x^2 + 3x - 15 = 0$$

A. 2

C. 5

B. 4

D. 6

56) If $f(x) = 5^x$ and $g(x) = \log_5 x$, which of the following expressions is equal to $f(5g(p))$?

A. $5P$

C. p^5

B. 5^p

D. $\frac{p}{5}$

57) The cost of using a car is $0.32 per minutes. Which of the following equations represents the total cost c, in dollars, for h hours of using the car?

A. $c = \frac{60h}{0.32}$

C. $c = \frac{0.32}{60h}$

B. $c = 0.32\,(60h)$

D. $c = 0.32h + 60$

58) Mary's average score after 5 tests is 90. What score on the 6th test would bring Mary's average up to exactly 91?

A. 96

C. 94

B. 98

D. 92

59) x is $y\%$ of what number?

A. $\frac{100y}{x}$

C. $\frac{x}{100y}$

B. $\frac{100x}{y}$

D. $\frac{y}{100x}$

60) For $i = \sqrt{-1}$, what is the value of $\frac{3+2i}{4+i}$?

A. $3 + i$

B. $\frac{32i}{5}$

C. $\frac{17-i}{5}$

D. $\frac{14+5i}{17}$

STOP

This is the End of this Section. You may check your work on this section if you still have time.

CLEP College Mathematics

Practice Test 2

❖ 60 Questions.

❖ Total time for this test: 90 Minutes.

❖ You may use a scientific calculator on this test.

Administered *Month Year*

1) If $xp + 3yq = 18$ and $xp + 2yq = 12$, what is the value of yq?

   A. 4

   B. 5

   C. 6

   D. 3

2) If $x^2 + 3$ and $x^2 - 3$ are two factors of the polynomial $8x^4 + n$ and n is a constant, what is the value of n?

   A. -72

   B. -36

   C. 36

   D. 72

3) If $3x - 2 = 7.6$, what is the value of $3x + 2$?

   A. 11.6

   B. 13.5

   C. 10.5

   D. 20

4) If the function f is defined by $f(x) = x^2 + 2x - 3$, which of the following is equivalent to $f(3t^2)$?

   A. $9t^4 + 18t^2 - 3$

   B. $9t^4 + 6t^2 - 3$

   C. $9t^4 + 3t^2 - 6t$

   D. $9t^4 + 8t^2 + 3$

5) The circle graph below shows all Mr. Green's expenses for last month. If he spent $680 on his car, how much did he spend for his rent?

   A. $750

   B. $760

   C. $716

   D. $816

   Mr. Green's monthly expenses

   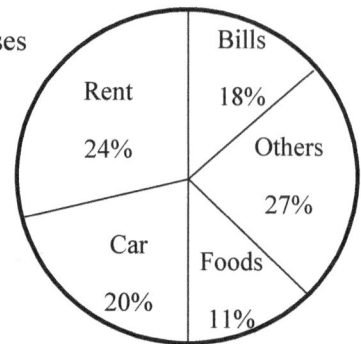

0.ABC   0.0D

6) The letters represent two decimals listed above. One of the decimals is equivalent to $\frac{3}{8}$ and the other is equivalent to $\frac{1}{20}$. What is the product of C and D?

A. 1

B. 10

C. 25

D. 15

7) The radius of circle A is three times the radius of circle B. If the circumference of circle A is 18π, what is the area of circle B?

A. 6π

B. 8π

C. 9π

D. 18π

8) In the diagram below, circle A represents the set of all odd numbers, circle B represents the set of all negative numbers, and circle C represents the set of all multiples of 5. Which number could be replaced with y?

A. 5

B. 0

C. −25

D. −20

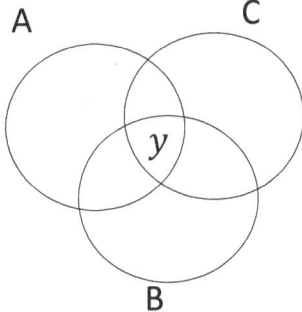

9) There are only red and blue cards in a box. The probability of choosing a red card in the box at random is half. If there are 126 blue cards, how many cards are in the box?

A. 138

B. 342

C. 356

D. 252

CLEP College Math Workbook

10) Both $(x = -1)$ and $(x = 2)$ are solutions for which of the following equations?

   I. $x^2 - 2x + 8 = 0$ III. $3x^2 - 3x - 6 = 0$

   II. $4x^2 - 4x = 8$

   A. II only   C. II and III

   B. I and II   D. I, II and III

11) In a certain bookshelf of a library, there are 60 biology books, 43 history books, and 77 language books. What is the ratio of the number of biology books to the total number of books in this bookshelf?

   A. $\frac{1}{5}$ C. $\frac{2}{5}$

   B. $\frac{1}{3}$ D. $\frac{1}{2}$

12) A container holds 1.5 gallons of water when it is $\frac{7}{42}$ full. How many gallons of water does the container hold when it's full?

   A. 7   C. 11

   B. 9   D. 14

13) The following table represents the value of x and function $f(x)$. Which of the following could be the equation of the function $f(x)$?

   A. $f(x) = x^2 - 6$ C. $f(x) = \sqrt{x+2}$

   B. $f(x) = x^2 - 5$ D. $f(x) = 2\sqrt{x} + 2$

x	$f(x)$
1	4
4	6
9	8
16	10

14) If 14% of x is 49 and $\frac{1}{4}$ of y is 18, what is the value of $x - y$?

  A. 278

  B. 350

  C. 72

  D. 31

15) Michelle and Alec can finish a job together in 40 minutes. If Michelle can do the job by herself in 4 hours, how many minutes does it take Alec to finish the job?

  A. 56

  B. 44

  C. 48

  D. 52

16) Angle a is 810 degrees and can be written $x\pi$ in radian. What is the value of x?

  A. 2.25

  B. 4.5

  C. 9

  D. 18

17) Which of the following expressions is equal to $\sqrt{\frac{x^2}{3} + \frac{x^2}{9}}$?

  A. x

  B. $\frac{2x}{3}$

  C. $x\sqrt{x}$

  D. $\frac{x\sqrt{x}}{3}$

18) In the following figure, point O is the center of the circle and the equilateral triangle has perimeter 42. What is the circumference of the circle? ($\pi = 3$)

  A. 42

  B. 21

  C. 84

  D. 96

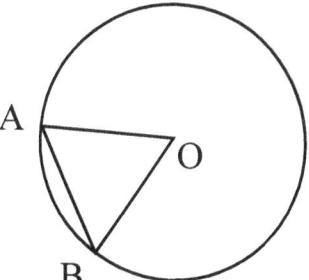

19) If $(3^a)^b = 81$, then what is the value of ab?

A. 5

B. 9

C. 4

D. 6

20) What is the sum of $\sqrt{x} - 11$ and $\sqrt{x} - 11$ when $\sqrt{x} = 6$?

A. -2

B. -4

C. 0

D. 2

21) What is the average (arithmetic mean) of all integers from 11 to 17?

A. 14

B. 14.5

C. 15

D. 15.5

22) What is the value of $\frac{4a-2}{5}$, if $-2a + 6a + 4a = 64$?

A. 5.5

B. 5

C. 6

D. 6.5

23) What is the value of $|-14 - 6| - |-10 + 4|$?

A. -24

B. 14

C. 24

D. -14

24) In the figure below, what is the value of x?

A. 57

B. 77

C. 69

D. 115

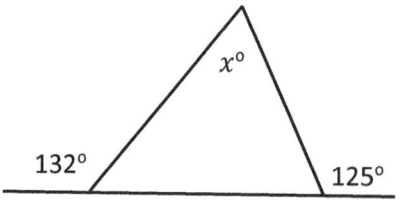

25) The table represents different values of function $g(x)$. What is the value of $5g(-1) - 3g(3)$?

x	-2	-1	0	2	3	4
$g(x)$	4	3	2	0	-1	-3

A. -18

B. -1

C. 3

D. 18

26) f a is an odd integer divisible by 7. Which of the following must be divisible by 3?

A. $a - 2$

B. $a + 2$

C. $2a$

D. $3a - 3$

27) If $(x - 2)^2 = 16$ which of the following could be the value of $(x - 3)(x - 4)$?

A. 5

B. 6

C. -5

D. -6

28) On the following figure, what is the area of the quadrilateral ABCD?

A. 28

B. 12

C. 24

D. 36

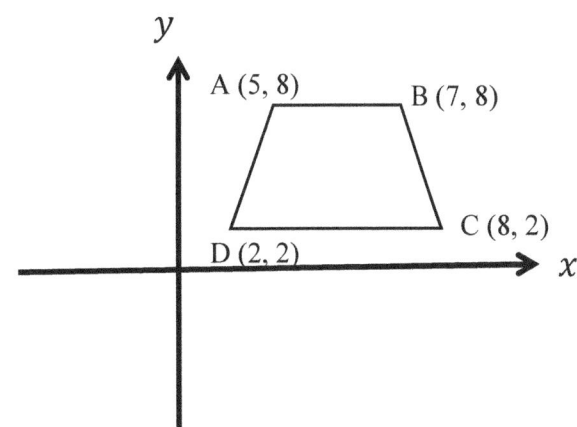

Questions 11 to 13 are base

29) Between which two of the months shown was there a twenty percent increase in the number of pants sold?

   A. January and February

   B. February and March

   C. March and April

   D. April and May

30) During the six-month period shown, what is the mean number of shirts and median number of shoes per month?

   A. 30, 137.5

   B. 149, 25

   C. 139, 25

   D. 20, 139

31) How many shoes need to be added in February until the ratio of number of pants to number of shoes in February equals to seven-eighth of this ratio in March?

   A. 2

   B. 9

   C. 5

   D. 18

CLEP College Math Workbook

32) What is the x-intercept of the line with equation $4x - 4y = 9$?

    A. -5

    B. -3

    C. $\frac{9}{4}$

    D. $\frac{5}{4}$

33) The base of a right triangle is 6 feet, and the interior angles are 45-45-90. What is its area?

    A. 18 square feet

    B. 16 square feet

    C. 12 square feet

    D. 36 square feet

34) In 1999, the average worker's income increased $2,000 per year starting from $24,000 annual salary. Which equation represents income greater than average? (I = income, x = number of years after 1999)

    A. $I > 2,000\ x + 24,000$

    B. $I > -2,000\ x + 24,000$

    C. $I < -2,000\ x + 24,000$

    D. $I < 2,000\ x - 24,000$

35) The Jackson Library is ordering some bookshelves. If x is the number of bookshelves the library wants to order, which each costs $300 and there is a one-time delivery charge of $700, which of the following represents the total cost, in dollar, per bookshelf?

    A. $300x + 700$

    B. $300 + 700x$

    C. $\frac{300x+700}{300}$

    D. $\frac{300x+700}{x}$

A library has 550 books that include Mathematics, Physics, Chemistry, English and History.

Use following graph to answer questions 18 to 20.

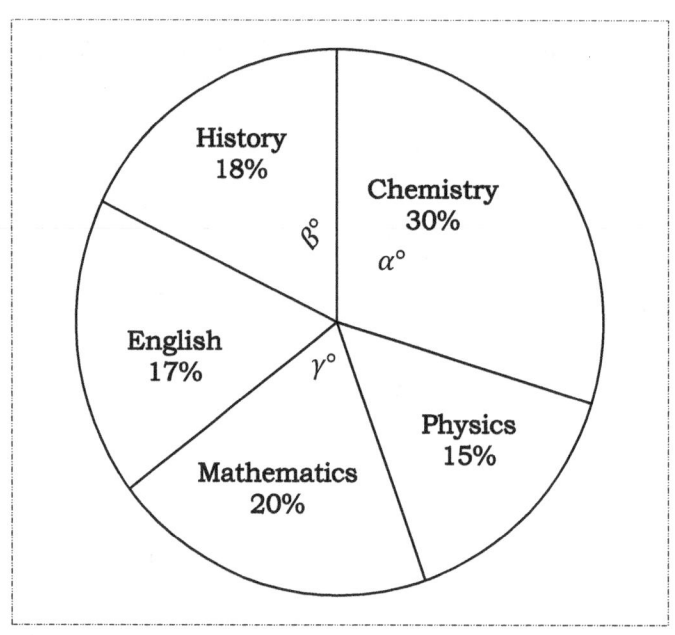

36) What is the product of the number of Mathematics and number of Chemistry books?

A. 18,150

B. 21,752

C. 28,460

D. 16,640

37) The librarians decided to move some of the books in the Chemistry section to Mathematics section. How many books are in the Mathematics section if now $\gamma = \frac{2}{3}\alpha$?

A. 50

B. 220

C. 140

D. 110

38) What are the values of angle γ and β respectively?

   A. 80°, 36°

   B. 125°, 36°

   C. 72°, 46.5°

   D. 72°, 64.3°

39) In the following figure, point Q lies online n, what is the value of y if $x = 20$?

   A. 10

   B. 20

   C. 55

   D. 35

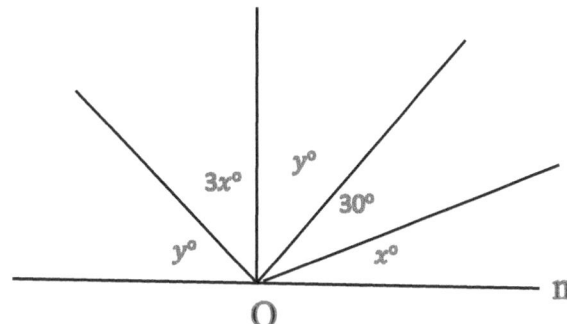

40) In the following figure, AB is the diameter of the circle. What is the circumference of the circle?

   A. 10 π

   B. 6 π

   C. 24 π

   D. 12 π

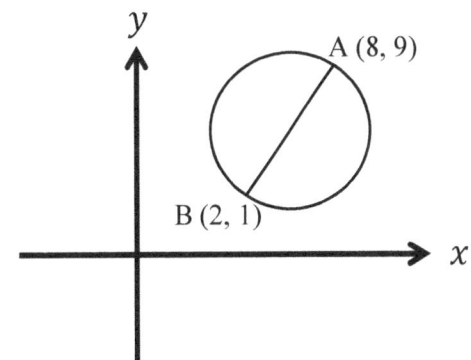

41) What is the smallest integer whose square root is greater than 6?

   A. 25

   B. 4

   C. 35

   D. 49

42) If the area of trapezoid is 30 cm, what is the perimeter of the trapezoid?

   A. 20 cm

   B. 22 cm

   C. 36 cm

   D. 50 cm

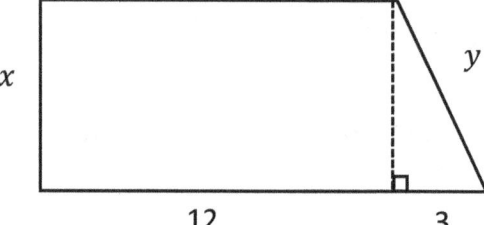

43) What is the solution of the following inequality?

$$|x - 2| \geq 8$$

   A. $x \geq 10 \cup x \leq -6$ C. $x \geq 10$

   B. $-6 \leq x \leq 10$ D. $x \leq -6$

44) If the area of the following rectangular ABCD is 160, and E is the midpoint of AB, what is the area of the shaded part?

   A. 20

   B. 80

   C. 50

   D. 120

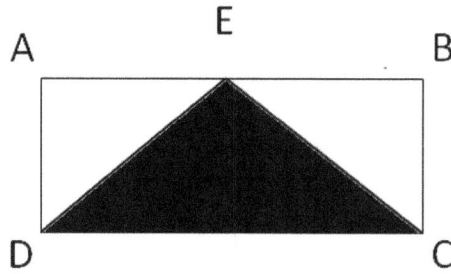

45) Which of the following is equivalent to $6 < -5x - 4 < 11$?

   A. $-3 < x < -2$ C. $2 < x < 5$

   B. $2 < x < 3$ D. $\frac{-3}{8} < x < \frac{1}{5}$

46) What are the zeroes of the function $f(x) = x^3 + 6x^2 + 8x$?

A. $-2, -4$

B. $0, 2, 4$

C. $-1, -4$

D. $0, -2, -4$

47) If m is a positive integer and $\sqrt{3m+54} = m$, what is the value of m?

A. 9

B. 18

C. 24

D. 54

48) What is the equation of the graph?

A. $x^2 + 6x + 5$

B. $x^2 + 2x + 4$

C. $2x^2 - 4x + 4$

D. $2x^2 + 4x + 2$

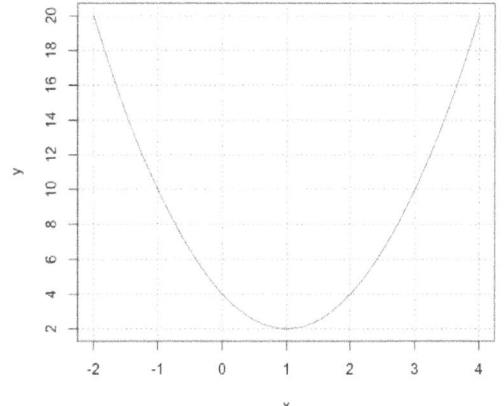

49) When 40% of 70 is added to 15% of 600, the resulting number is:

A. 90

B. 118

C. 128

D. 185

50) $(x^5)^{\frac{3}{10}}$ equal to?

A. $x\sqrt{x}$

B. $\sqrt[10]{x^{\frac{3}{5}}}$

C. $\sqrt[3]{x^2}$

D. $\sqrt{x^{\frac{3}{2}}}$

51) In the following figure, ABCD is a rectangle. If $a = \sqrt{2}$, and $b = 2a$, find the area of the shaded region? (The shaded region is a trapezoid)

A. $\sqrt{2}$

B. $3\sqrt{2}$

C. $2\sqrt{3}$

D. $4\sqrt{2}$

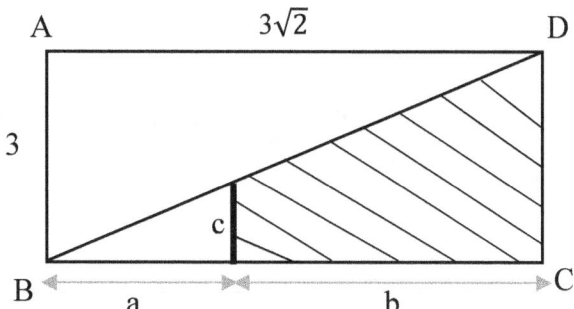

52) 3 liters of water are poured into an aquarium that's 15 cm long, 4 cm wide, and 80 cm high. How many cm will the water level in the aquarium rise due to this added water? (1 liter of water = 1,000 cm^3)?

A. 60

B. 20

C. 50

D. 45

53) If $A = \{1, 3, 5, 8, 16, 24\}$ and $B = \{1, 3, 12, 24, 32\}$, how many elements are in $A \cup B$?

A. 2

B. 3

C. 6

D. 8

54) For $i = \sqrt{-1}$, which of the following is equivalent of $\frac{2+3i}{4-2i}$?

A. $\frac{3+2i}{5}$

B. $\frac{2+16i}{20}$

C. $\frac{2+16i}{12}$

D. $\frac{14+16i}{12}$

55) Convert 350,000 to scientific notation.

   A. 3.50×1000

   B. 3.50×10^{-5}

   C. 3.5×10^5

   D. 3.5×10^4

56) If $\alpha = 2\beta$ and $\beta = 4\gamma$, how many α are equal to 64γ?

   A. 10

   B. 8

   C. 2

   D. 1

57) If $f(x) = \frac{6x-3}{5}$ and $f^{-1}(x)$, is the inverse of $f(x)$, what is the value of $f^{-1}(3)$?

   A. 5

   B. $\frac{1}{3}$

   C. $\frac{1}{5}$

   D. 3

58) Sara orders a box of pen for $4 per box. A tax of 8.5% is added to the cost of the pens before a flat shipping fee of $8 closest out the transaction. Which of the following represents total cost of p boxes of pens in dollars?

   A. $8p + 4$

   B. $1.085(4p) + 8$

   C. $1.085(8p) + 4$

   D. $4p + 8$

59) A construction company is building a wall. The company can build 20 cm of the wall per minute. After 30 minutes $\frac{3}{2}$ of the wall is completed. How many meters is the wall?

   A. 6

   B. 8

   C. 4

   D. 16

60) If $f(x)=3x^3+3$ and $g(x)=\frac{1}{x}$, what is the value of $f(g(x))$?

A. $\frac{1}{3x^3+3}$

B. $\frac{1}{3x}$

C. $\frac{1}{3x+3}$

D. $\frac{3}{x^3}+3$

STOP

This is the End of this Section. You may check your work on this section if you still have time.

Answers and Explanations

CLEP College Mathematics Practice Tests

Answer Key

❋ Now, it's time to review your results to see where you went wrong and what areas you need to improve!

CLEP College Mathematics Practice Tests

Practice Test 1

1	A	21	B	41	D
2	B	22	D	42	B
3	B	23	C	43	C
4	A	24	A	44	C
5	C	25	C	45	D
6	A	26	A	46	A
7	B	27	C	47	D
8	D	28	B	48	A
9	C	29	D	49	B
10	B	30	A	50	A
11	B	31	B	51	C
12	D	32	B	52	C
13	B	33	C	53	B
14	C	34	D	54	B
15	A	35	C	55	C
16	B	36	B	56	C
17	D	37	C	57	B
18	D	38	C	58	A
19	C	39	D	59	B
20	C	40	B	60	D

Practice Test 2

1	C	21	A	41	D
2	A	22	C	42	C
3	A	23	B	43	A
4	B	24	B	44	B
5	D	25	D	45	A
6	C	26	D	46	D
7	C	27	B	47	A
8	C	28	C	48	C
9	D	29	D	49	B
10	C	30	C	50	A
11	B	31	B	51	D
12	B	32	C	52	C
13	D	33	A	53	D
14	A	34	A	54	B
15	C	35	D	55	C
16	B	36	A	56	B
17	B	37	D	57	D
18	D	38	D	58	B
19	C	39	D	59	C
20	C	40	A	60	D

Practice Tests 1

Answers and Explanations

1) **Answer: A.**

$4a = 5b \to b = \frac{4a}{5}$ and $4a = 3c \to c = \frac{4a}{3}$

$2a + 5b + 6c = 2a + \left(5 \times \frac{4a}{5}\right) + \left(6 \times \frac{4a}{3}\right) = 2a + 4a + 8a = 14a$; The value of

$2a + 5b + 6c$ is 14 times the value of a.

2) **Answer: B.**

$x^2 = 144 \to x = 12$ (positive value) Or $x = -12$ (negative value)

Since x is positive, then:

$f(144) = f(12^2) = 6(12) + 4 = 72 + 4 = 76$

3) **Answer: B.**

$3x^2 - 22x + 16 = -7x + 34 \to 3x^2 - 22x + 7x + 16 - 34 = 0$

$\to 3x^2 - 15x - 18 = 0 \to 3(x^2 - 5x - 6) = 0 \to$ Divide both sides by 3. Then:

$x^2 - 5x - 6 = 0$, Find the factors of the quadratic equation.

$\to (x - 6)(x + 1) = 0 \to x = 6$ or $x = -1$

$a > b$, then: $a = 6$ and $b = -1 \to \frac{a}{b} = \frac{6}{-1} = -6$

4) **Answer: A.**

To rewrite $\frac{1}{\frac{1}{x-6}+\frac{1}{x+7}}$, first simplify $\frac{1}{x-6} + \frac{1}{x+7}$.

$\frac{1}{x-6} + \frac{1}{x+7} = \frac{(x+7)}{(x-6)(x+7)} + \frac{(x-6)}{(x+7)(x-6)} = \frac{(x+7)+(x-6)}{(x+7)(x-6)}$

Then: $\frac{1}{\frac{1}{x-6}+\frac{1}{x+7}} = \frac{1}{\frac{(x+7)+(x-6)}{(x+7)(x-6)}} = \frac{(x-6)(x+7)}{(x-6)+(x+7)}$. (Remember, $\frac{1}{\frac{1}{x}} = x$)

This result is equivalent to the expression in choice A.

5) **Answer: C.**

First, find the equation of the line. All lines through the origin are of the form $y = mx$, so the equation is $y = \frac{1}{2}x$. Of the given choices, only choice C (4,2), satisfies this equation: $y = \frac{1}{2}x \to 2 = \frac{1}{2}(4) = 2$

CLEP College Math Workbook

6) Answer: A.

$4x + 4.5 > 12x - 9.5 - 4.5x \rightarrow$ Combine like terms:

$4x + 4.5 > 7.5x - 9.5 \rightarrow$ Subtract $4x$ from both sides:

$4.5 > 3.5x - 9.5$

Add 9.5 both sides of the inequality.

$14 > 3.5x$, Divide both sides by 3.5: $\frac{14}{3.5} > x \rightarrow x < 4$

7) Answer: B.

Of the 40 employees, there are 13 females under age 55 and 11 males age 55 or older. Therefore, the probability that the person selected will be either a female under age 55 or a male age 55 or older is: $\frac{13}{40} + \frac{11}{40} = \frac{24}{40} = \frac{3}{5}$

8) Answer: D.

Plug in the values of x and y of the point $(3, 7)$ in the equation of the parabola. Then:

$7 = a(3)^2 + 4(3) + 22 \rightarrow 7 = 9a + 12 + 22 \rightarrow 7 = 9a + 34$

$\rightarrow 9a = 7 - 34 = -27 \rightarrow a = \frac{-27}{9} = -3 \rightarrow a^2 = (-3)^2 = 9$

9) Answer: C.

x is the number of all John's sales per month and 8% of it is: $8\% \times x = 0.08x$

John's monthly revenue: $0.08x + 3,400$

10) Answer: B.

The output value is 0. Then: $x = 3$

$f(x) = x^3 - 9x \rightarrow f(3) = 3^3 - 9(3) = 27 - 27 = 0$

11) Answer: B.

To solve this problem, first recall the equation of a line: $y = mx + b$; Where, $m = slope$ and $y = y - intercept$

Remember that slope is the rate of change that occurs in a function and that the $y-$intercept is the y value corresponding to $x = 0$.

Since the height of John's plant is 7 inches tall when he gets it. Time (or x) is zero. The plant grows 5 inches per year. Therefore, the rate of change of the plant's height is 5. The $y-$intercept represents the starting height of the plant which is 7 inches.

WWW.MathNotion.Com

12) Answer: D.

Two triangles ΔBAE and ΔBCD are similar. Then:

$\frac{AE}{CD} = \frac{AB}{BC} \to \frac{3}{2} = \frac{x}{15-x} \to 3(15-x) = 2x \to 2x + 3x = 3 \times 15 \to 5x = 45 \to x = 9$

13) Answer: B.

$\begin{cases} \frac{-x}{6} + \frac{y}{3} = 2 \\ \frac{-6}{8}x + 2y = 12 \end{cases}$ Multiply the top equation by -6. Then, $\begin{cases} x - 2y = -12 \\ \frac{-6}{8}x + 2y = 12 \end{cases}$ Add two equations.

$\frac{2}{8}x = 0 \to x = 0$, plug in the value of x into the first equation $\to y = 6$

14) Answer: C.

First, simplify the numerator and the denominator.

$\frac{(9(x^2)(y^3))^3}{(6xy^2)^3} = \frac{729x^6y^9}{216x^3y^6} = \frac{729x^6y^9}{216x^3y^6} = \frac{27}{8}x^{6-3}y^{9-6} = \frac{27}{8}x^3y^3 = (\frac{3}{2}xy)^3$

15) Answer: A.

$\frac{y}{8} = x - \frac{3}{8}x + 9$, Multiply both sides of the equation by 8. Then:

$8 \times \frac{y}{8} = 8 \times (x - \frac{3}{8}x + 9) \to y = 8x - 3x + 72 \to y = 5x + 72$

Now, subtract $4x$ from both sides of the equation. Then, $y - 5x = 72$

16) Answer: B.

Let L be the length of the rectangular and W be the with of the rectangular.

Then, $L = 7W + 2$; The perimeter of the rectangle is 36 meters. Therefore:

$2L + 2W = 36 \Rightarrow L + W = 18$

Replace the value of L from the first equation into the second equation and solve for

$W: (7W + 2) + W = 18 \to 8W + 2 = 18 \to 8W = 16 \to W = 2$

The width of the rectangle is 2 meters and its length is: $L = 7W + 2 = 7(2) + 2 = 16$

The area of the rectangle is: length × width = 2 × 16 = 32

17) Answer: D.

Substituting 8 for x and 12 for y in $y = nx + 5$ gives $12 = (n)(8) + 5$

which gives $n = 1$. Hence, $y = x + 5$. Therefore, when $x = 8$, the value of y is:

$$y = 8 + 5 = 13$$

18) Answer: D.

The description $6 + 3x$ is 12 more than 18 can be written as the equation $6 + 3x = 12 + 18$, which is equivalent to $6 + 3x = 30$. Subtracting 6 from each side of $6 + 3x = 30$ gives

$3x = 24$. Since $9x$ is 3 times $3x$, multiplying both sides of $3x = 24$ by 3 gives $9x = 72$

19) Answer: C.

$\frac{4}{5} \times 30 = \frac{120}{5} = 24$

20) Answer: C.

The slop of line A is: $m = \frac{y_2 - y_1}{x_2 - x_1} = \frac{7-6}{3-2} = 1$

Parallel lines have the same slope and only choice D ($y = x$) has slope of 1.

21) Answer: B.

Choices A, C and D are incorrect because 60% of each of the numbers is a non-whole number.

  A. 42, $60\% \; of \; 42 = 0.60 \times 42 = 25.2$
  B. 40, $60\% \; of \; 40 = 0.60 \times 40 = 24$
  C. 46, $60\% \; of \; 46 = 0.60 \times 46 = 27.6$
  D. 29, $60\% \; of \; 29 = 0.60 \times 29 = 17.4$

Only choice B gives a whole number.

22) Answer: D.

The capacity of a red box is 40% bigger than the capacity of a blue box and it can hold 35 books. Therefore, we want to find a number that 40% bigger than that number is 35. Let x be that number. Then: $1.40 \times x = 35$, Divide both sides of the equation by 1.4. Then: $x = \frac{35}{1.40} = 25$

23) Answer: C.

The smallest number is -15. To find the largest possible value of one of the other four integers, we need to choose the smallest possible integers for three of them. Let x be the largest number. Then: $-65 = (-15) + (-14) + (-13) + (-12) + x \rightarrow -65 = -54 + x \rightarrow x = -65 + 54 = -11$

24) Answer: A.

Let x be equal to 1.5, then: $x = 1.5$

$\sqrt{x^2 + 1} = \sqrt{1.5^2 + 1} = \sqrt{3.25} \approx 1.8$

$\sqrt{x^2} + 1 = \sqrt{1.5^2} + 1 = 1.5 + 1 = 2.5$

Then, option A is correct: $x < \sqrt{x^2 + 1} < \sqrt{x^2} + 1$

25) Answer: C.

The ratio of boy to girls is 3:7. Therefore, there are 3 boys out of 10 students. To find the answer, first divide the total number of students by 10, then multiply the result by 3.

$90 \div 10 = 9 \Rightarrow 3 \times 9 = 27$

There are 27 boys and 63 (90 – 27) girls. So, 36 more boys should be enrolled to make the ratio 1:1

26) Answer: A.

If $f(x) = 4x + 2(x + 2) + 2$, then find $f(3x)$ by substituting $3x$ for every x in the function. This gives: $f(3x) = 4(3x) + 2(3x + 2) + 2$,

It simplifies to: $f(4x) = 12x + 6x + 4 + 2 = 18x + 6$

27) Answer: C.

Let's find the mean (average), mode and median of the number of cities for each type of pollution.

Number of cities for each type of pollution: 5, 2, 3, 8, 7

$average\ (mean) = \frac{sum\ of\ terms}{number\ of\ terms} = \frac{5+2+3+8+7}{5} = \frac{25}{5} = 5$

Median is the number in the middle. To find median, first list numbers in order from smallest to largest: 2, 3, 5, 7, 8. Median of the data is 5.

Mode is the number which appears most often in a set of numbers. Therefore, there is no mode in the set of numbers.

Median = Mean, then, $a=c$

CLEP College Math Workbook

28) Answer: B.

Let the number of cities should be added to type of pollutions C be x. Then: $\frac{x+2}{8} = 0.50 \rightarrow x + 2 = 8 \times 0.50 \rightarrow x + 2 = 4 \rightarrow x = 2$

29) Answer: D.

Percent of cities in the type of pollution B: $\frac{3}{10} \times 100 = 30\%$

Percent of cities in the type of pollution C: $\frac{2}{10} \times 100 = 20\%$

Percent of cities in the type of pollution E: $\frac{5}{10} \times 100 = 50\%$

30) Answer: A.

$AB = 3$ And $BC = 4$

$AC = \sqrt{4^2 + 3^2} = \sqrt{16 + 9} = \sqrt{25} = 5$

Perimeter $= 3 + 4 + 5 = 12$

Area $= \frac{3 \times 4}{2} = 3 \times 2 = 6$

In this case, the ratio of the perimeter of the triangle to its area is: $\frac{12}{6} = 2$

If the sides AB and BC become tripe longer, then: $AB = 9$ And $BC = 12$

$AC = \sqrt{9^2 + 12^2} = \sqrt{81 + 144} = \sqrt{225} = 15$

Perimeter $= 15 + 12 + 9 = 36$

Area $= \frac{9 \times 12}{2} = 9 \times 6 = 54$

In this case the ratio of the perimeter of the triangle to its area is: $\frac{36}{54} = \frac{2}{3}$

31) Answer: B.

Since $f(x)$ is linear function with a negative slop, then when $x = -1, f(x)$ is maximum and when $x = 2, f(x)$ is minimum. Then the ratio of the minimum value to the maximum value of the function is: $\frac{f(2)}{f(-1)} = \frac{-5(2)+2}{-5(-1)+2} = \frac{-8}{7}$

32) Answer: B.

The equation $\frac{a-b}{b} = \frac{5}{9}$ can be rewritten as $\frac{a}{b} - \frac{b}{b} = \frac{5}{9}$, from which it follows that $\frac{a}{b} - 1 = \frac{5}{9}$, or $\frac{a}{b} = \frac{5}{9} + 1 = \frac{14}{9}$.

33) Answer: C.

First, factorize the numerator and simplify.

$\frac{x^2-16}{x+4} + 3(x+3) = 12 \rightarrow \frac{(x-4)(x+4)}{x+4} + 3x + 9 = 12$

Divide both sides of the fraction by $(x+4)$. Then:

$x - 4 + 3x + 9 = 12 \rightarrow 4x + 5 = 12$

Subtract 5 from both sides of the equation.

Then: $4x = 12 - 5 = 7 \rightarrow x = \frac{7}{4}$

34) Answer: D.

Percentage of men in city C = $\frac{710}{1,385} \times 100 = 51.26\%$

Percentage of men in city B = $\frac{320}{631} \times 100 = 50.71\%$

Percentage of men in city C to percentage of men in city B: $\frac{51.26}{50.71} = 99.11$

35) Answer: C.

Ratio of women to men in city A: $\frac{520}{550} = 0.95$

Ratio of women to men in city B: $\frac{311}{320} = 0.97$

Ratio of women to men in city C: $\frac{675}{710} = 0.95$

Ratio of women to men in city D: $\frac{548}{570} = 0.96$

36) Answer: B.

Let the number of women should be added to city D be x, then:

$\frac{548 + x}{570} = 1.4 \rightarrow 548 + x = 570 \times 1.4 = 798 \rightarrow x = 228$

37) Answer: C.

The perimeter of the rectangle is: $2x + 2y = 24 \rightarrow x + y = 12 \rightarrow x = 12 - y$

The area of the rectangle is: $x \times y = 32 \rightarrow (12 - y)(y) = 32 \rightarrow y^2 - 12y + 32 = 0$

Solve the quadratic equation by factoring method. $(y - 4)(y - 8) = 0$

$y = 4$ (Unacceptable, because y must be greater than 4)

or $y = 8 \rightarrow x \times y = 32 \rightarrow x \times 8 = 32 \rightarrow x = 4$

38) Answer: C.

The amount of petrol consumed after x hours is: $5 \times x = 5x$

Petrol remaining after x hours driving: $80 - 5x$

39) Answer: D.

In the figure angle A is labeled $(3x + 5)$ and it measures 35. Thus, $3x + 5 = 35$ and $3x = 30$ or $x = 10$.

That means that angle B, which is labeled $(7x)$, must measure $7 \times 10 = 70$.

Since the three angles of a triangle must add up to 180, $35 + 70 + y - 8 = 180$, then: $y + 97 = 180 \rightarrow y = 180 - 97 = 83$

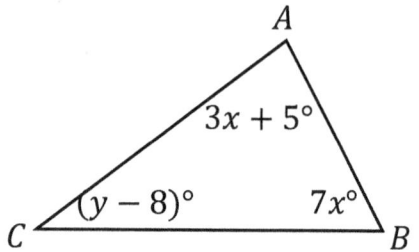

40) Answer: B.

$average\ (mean) = \dfrac{sum\ of\ terms}{number\ of\ terms} = \dfrac{7+11+13+15+18.5+14+12.5}{7} = 13$

41) Answer: D.

$\begin{cases} x + 3y = 2 \\ -3x - 2y = 8 \end{cases} \rightarrow$ Multiply the top equation by 3 then,

$\begin{cases} 3x + 9y = 6 \\ -3x - 2y = 8 \end{cases} \rightarrow$ Add two equations

$7y = 14 \rightarrow y = 2$, plug in the value of y into the first equation:

$x + 3y = 2 \rightarrow x + 3(2) = 2 \rightarrow x + 6 = 2$

Add -6 from both sides of the equation.

Then: $x + 6 - 6 = 2 - 6 \rightarrow x = -4$

42) Answer: B.

$\sin \beta = \dfrac{Opposite\ side}{hypotenuse}$

To find the hypotenuse, we need to use Pythagorean theorem.

$a^2 + b^2 = c^2 \rightarrow c = \sqrt{a^2 + b^2}$

$\sin(\beta) = \dfrac{b}{c} = \dfrac{b}{\sqrt{a^2 + b^2}}$

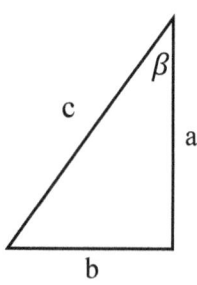

43) Answer: C.

$\left|\frac{x}{3} - x + 8\right| < 2 \to \left|-\frac{2}{3}x + 8\right| < 2 \to -2 < -\frac{2}{3}x + 8 < 2$

Subtract 8 from all sides of the inequality.

$\to -2 - 8 < -\frac{2}{3}x + 8 - 8 < 2 - 8 \to -10 < -\frac{2}{3}x < -6$

Multiply all sides by 3.

$\to 3 \times (-10) < 3 \times \left(-\frac{2x}{3}\right) < 3 \times (-6) \to -30 < -2x < -18$

Divide all sides by -2. (Remember that when you divide all sides of an inequality by a negative number, the inequality sing will be swapped. $<$ becomes $>$)

$\to \frac{-30}{-2} > \frac{-2x}{-2} > \frac{-18}{-2} \to 15 > x > 9 \to 9 < x < 15$

44) Answer: C.

x is directly proportional to the square of y. Then: $x = cy^2$

$36 = c(3)^2 \to 36 = 9c \to c = \frac{36}{9} = 4$

The relationship between x and y is: $x = 4y^2$

$x = 256 \to 256 = 4y^2 \to y^2 = \frac{256}{4} = 64 \to y = 8$

45) Answer: D.

The union of A and B is: $A \cup B = \{1, 2, 3, 4, 8, 6, 11, 10\}$

The intersection of $(A \cup B)$ and C is: $(A \cup B) \cap C = \{8, 10\}$

46) Answer: A.

The intersection of two functions is the point with 3 for x. Then:

$f(2) = g(2)$ and $g(2) = (3 \times (2)) - 2 = 6 - 2 = 4$

Then, $f(2) = 4 \to a(2)^2 + b(2) + c = 4 \to 4a + 2b + c = 4$ (i)

The value of x in the vertex of the parabola is:

$x = -\frac{b}{2a} \to -2 = -\frac{b}{2a} \to b = 4a$ (ii)

In the point $(-2, 0)$, $f(-2) = 0 \to a(-2)^2 + b(-2) + c = 0$

$\to 4a - 2b + c = 0$ (iii)

Using the first two equation: $\begin{cases} 4a + 2b + c = 4 \\ 4a - 2b + c = 0 \end{cases}$

Equation 1 minus equation 2 is: (i)−(iii) → $4b = 4$ → $b = 1$ (iv)

Plug in the value of b in the second equation: $b = 4a$ → $a = \frac{b}{4} = \frac{1}{4} = \frac{1}{4}$

Plug in the values of a and b in the first equation. Then:

$4\left(\frac{1}{4}\right) + 2(1) + c = 4 \to 1 + 2 + c = 4 \to c = 4 - 3 \to c = 1$

The product of a, b and c =$\left(\frac{1}{4}\right) \times 1 \times 1 = \frac{1}{4}$

47) Answer: D.

The relationship among all sides of special right triangle $30°, 60°, 90°$ is provided in this triangle:

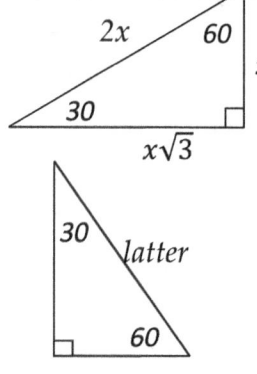

In this triangle, the opposite side of $30°$ angle is half of the hypotenuse.

Draw the shape of this question.

The latter is the hypotenuse. Therefore, the latter is 48 feet.

48) Answer: A.

The sum of supplement angles is 180. Let x be that angle. Therefore, $x + 9x = 180 \Rightarrow 10x = 180$, divide both sides by 10: $x = 18$

49) Answer: B.

$3x + 6y = \frac{-3y^2 + 18}{x}$, Multiply both sides by x.

$x \times (3x + 6y) = x \times \left(\frac{-3y^2 + 18}{x}\right) \to 3x^2 + 6xy = -3y^2 + 18$

$\to 3x^2 + 6xy + 3y^2 = 18 \to 3 \times (x^2 + 2xy + y^2) = 18 \to x^2 + 2xy + y^2 = \frac{18}{3}$

$x^2 + 2xy + y^2 = (x + y)^2$, Then: $(x + y)^2 = 6$

50) Answer: A.

Let x be the length of an edge of cube, then the volume of a cube is: $V = x^3$

The surface area of cube is: $SA = 6x^2$

The volume of cube A is $\frac{1}{3}$ of its surface area. Then, $x^3 = \frac{6x^2}{3}$

$\to x^3 = 2x^2$, divide both side of the equation by $x^2 \to \frac{x^3}{x^2} = \frac{2x^2}{x^2} \to x = 2$

51) Answer: C.

First square both sides of the equation to get $3m - 2 = m^2$

Subtracting both sides by $3m - 2$ gives us the equation $m^2 - 3m + 2 = 0$

Here you can solve the quadratic equation by factoring to get $(m - 1)(m - 2) = 0$

For the expression $(m - 1)(m - 2)$ to equal zero, $m = 1$ or $m = 2$

52) Answer: C.

A zero of a function corresponds to an x-intercept of the graph of the function in the xy-plane. Therefore, the graph of the function $g(x)$, which has three distinct zeros, must have three x−intercepts. Only the graph in choice C has three x−intercepts.

53) Answer: B.

Replace z by $z/5$ and simplify.

$$x_1 = \frac{8y + \frac{r}{r+1}}{\frac{6}{\frac{z}{5}}} = \frac{8y + \frac{r}{r+1}}{\frac{5 \times 6}{z}} = \frac{8y + \frac{r}{r+1}}{5 \times \frac{6}{z}} = \frac{1}{5} \times \frac{8y + \frac{r}{r+1}}{\frac{6}{z}} = \frac{x}{5}$$

When z is divided by 5, x is also divided by 5.

54) Answer: B.

Use the information provided in the question to draw the shape.

Use Pythagorean Theorem: $a^2 + b^2 = c^2$

$150^2 + 80^2 = c^2 \Rightarrow 22500 + 6400 = c^2$

$\Rightarrow 28900 = c^2 \Rightarrow c = 170$

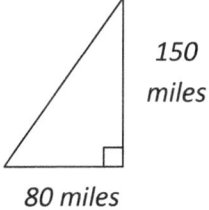

55) Answer: C.

The four-term polynomial expression can be factored completely, by grouping, as follows: $(x^3 - 5x^2) + (3x - 15) = 0$

$x^2(x - 5) + 3(x - 5) = 0 \quad (x - 5)(x^2 + 3) = 0$

By the zero-product property, set each factor of the polynomial equal to 0 and solve each resulting equation for x. This gives $x = 5 \: or \: x = \pm i\sqrt{3}$, respectively. Because the equation the question asks for the real value of x that satisfies the equation, the correct answer is 5.

CLEP College Math Workbook

56) Answer: C.

To solve for $f(5g(P))$, first, find $5g(p)$. $g(x) = log_5 x \to g(p) = log_5 p \to 3g(p) = 5 log_5 p = log_5 p^5$

Now, find $f(5g(p))$: $f(x) = 5^x \to f(log_5 p^5) = 5^{log_5 p^5}$

Logarithms and exponentials with the same base cancel each other. This is true because logarithms and exponentials are inverse operations. Then: $f(log_5 p^5) = 5^{log_5 p^5} = p^5$

57) Answer: B.

$0.32 per minute to use car. This per-minute rate can be converted to the hourly rate using the conversion 1 hour = 60 minutes, as shown below.

$\frac{0.32}{minute} \times \frac{60 \ minutes}{1 \ hours} = \frac{\$(0.32 \times 60)}{hour}$. Thus, the car costs $\$(0.32 \times 60)$ per hour.

Therefore, the cost c, in dollars, for h hours of use is $c = (0.32 \times 60)h$,

Which is equivalent to $c = 0.32(60h)$

58) Answer: A.

The best way to deal with changing averages is to use the sum. Use the old average to figure out the total of the first 5 scores: Sum of first 5 scores: $(5)(90) = 450$

Use the new average to figure out the total she needs after the 6th score:

Sum of 6 score: $(6)(91) = 546$

To get her sum from 450 to 552, Mary needs to score $5546 - 450 = 96$.

59) Answer: B.

Let the number be A. Then: $x = y\% \times A$. Solve for A. $x = \frac{y}{100} \times A$

Multiply both sides by $\frac{100}{y}$: $x \times \frac{100}{y} = \frac{y}{100} \times \frac{100}{y} \times A \to A = \frac{100x}{y}$

60) Answer: D.

To perform the division $\frac{3+2i}{4+i}$, multiply the numerator and denominator of $\frac{3+2i}{4+1i}$ by the conjugate of the denominator, $5 - i$. This gives $\frac{(3+2i)(4-i)}{(4+i)(4-i)} = \frac{12-3i+8i-2i^2}{4^2-i^2}$. Since $i^2 = -1$, this can be simplified to $\frac{12-3i+8i+2}{16+1} = \frac{14+5i}{17}$

Practice Tests 2
Answers and Explanations

1) Answer: C.

$xp + 3yq = 18 \to xp = 18 - 3yq$ (1)

$xp + 2yq = 12$ (2)

(1) in (2) $\to 18 - 3yq + 2yq = 12 \to 18 - yq = 12 \to yq = 18 - 12 = 6$

2) Answer: A.

$8x^4 + n = a(x^2 + 3)(x^2 - 3) = ax^4 - 9a \to a = 8$

And $n = -9a = -9 \times 8 = -72$

3) Answer: A.

$3x - 2 = 7.6 \to 3x = 7.6 + 2 = 9.6 \to x = \frac{9.6}{3} = 3.2$

Then; $3x + 2 = 3(3.2) + 2 = 9.6 + 2 = 11.6$

4) Answer: B.

$f(x) = x^2 + 2x - 3$

$f(3t^2) = (3t^2)^2 + 2(3t^2) - 3 = 9t^4 + 6t^2 - 3$

5) Answer: D.

Let x be all expenses, then $\frac{20}{100}x = \$680 \to x = \frac{100 \times \$680}{20} = \$3,400$

He spent for his rent: $\frac{24}{100} \times \$3,400 = \$816$

6) Answer: C.

$\frac{3}{8} = 0.375 \to C = 5$

$\frac{1}{20} = 0.05 \to D = 5 \to C \times D = 5 \times 5 = 25$

7) Answer: C.

Let P be circumference of circle A, then; $2\pi r_A = 18\pi \to r_A = 9$

$r_A = 3r_B \to r_B = \frac{9}{3} = 3 \to$ Area of circle B is $\pi r_B^2 = 9\pi$

8) Answer: C.

y is the intersection of the three circles. Therefore, it must be odd (from circle A), negative (from circle B), and multiple of 5 (from circle C).

From the options, only -25 is odd, negative and multiple of 5.

9) Answer: D.

let x be total number of cards in the box, then number of red cards is: $x - 126$

The probability of choosing a red card is one third. Then: probability $= \frac{1}{2} = \frac{x-126}{x}$

Use cross multiplication to solve for x.

$x \times 1 = 2(x - 126) \to x = 2x - 252 \to 2x - x = 252 \to x = 252$

10) Answer: C.

Plug in the values of x in each equation and check.

I. $(-1)^2 - 2(-1) + 8 = 1 + 2 + 8 = 11 \neq 0$

$(2)^2 - 2(2) + 8 = 4 - 4 + 8 = 8 \neq 0$

II. $4(-1)^2 - 4(-1) = 4 + 4 = 8 \to 8 = 8$

$4(2)^2 - 4(2) = 16 - 8 = 8 \to 8 = 8$

III. $3(-1)^2 - 3(-1) - 6 = 3 + 3 - 6 = 0 = 0$

$3(2)^2 - 3(2) - 6 = 12 - 6 - 6 = 0 = 0$

Equations II and III are correct.

11) Answer: B.

Number of biology book: 60

Total number of books; $60 + 43 + 77 = 180$

the ratio of the number of biology books to the total number of books is: $\frac{60}{180} = \frac{1}{3}$

12) Answer: B.

let x be the number of gallons of water the container holds when it is full.

Then; $\frac{7}{42}x = 1.5 \to x = \frac{42 \times 1.5}{7} = 9$

13) Answer: D.

A. $f(x) = x^2 - 6$; if $x = 1 \to f(1) = (1)^2 - 6 = 1 - 6 = -5 \neq 4$

B. $f(x) = x^2 - 3$; if $x = 1 \to f(1) = (1)^2 - 5 = 1 - 5 = -4 \neq 4$

C. $f(x) = \sqrt{x+2}$ if $x = 1 \to f(1) = \sqrt{1+2} = \sqrt{3} \neq 4$

D. $f(x) = 2\sqrt{x} + 2$ if $x = 1 \to f(1) = 2\sqrt{1} + 2 = 4 = 4$ (This is correct)

14) Answer: A.

$\frac{14}{100}x = 49 \rightarrow x = \frac{49 \times 100}{14} = 350$

$\frac{1}{4}y = 18 \rightarrow y = 4 \times 18 = 72 \rightarrow x - y = 350 - 72 = 278$

15) Answer: C.

Let b be the amount of time Alec can do the job, then,

$\frac{1}{a} + \frac{1}{b} = \frac{1}{40} \rightarrow \frac{1}{240} + \frac{1}{b} = \frac{1}{40} \rightarrow \frac{1}{b} = \frac{1}{40} - \frac{1}{240} = \frac{5}{240} = \frac{1}{48}$

Then: $b = 48$ minutes

16) Answer: B.

One-degree equals $\frac{\pi}{180}$.

The angle α in radians is equal to the angle α in degrees times π constant divided by 180 degrees. Then: $1\ degree = \frac{\pi}{180}; \rightarrow 810\ degrees = \frac{810\pi}{180} = 4.5\pi$

$4.5\pi = x\pi \rightarrow x = 4.5$

17) Answer: B.

Simplify the expression.

$\sqrt{\frac{x^2}{3} + \frac{x^2}{9}} = \sqrt{\frac{3x^2}{9} + \frac{x^2}{9}} = \sqrt{\frac{4x^2}{9}} = \sqrt{\frac{4}{9}x^2} = \sqrt{\frac{4}{9}} \times \sqrt{x^2} = \frac{2}{3} \times x = \frac{2x}{3}$

18) Answer: D.

In the equilateral triangle if x is length of one side of triangle, then the perimeter of the triangle is $3x$. Then $3x = 42 \rightarrow x = 14$ and radius of the circle is: $x = 14$

Then, the perimeter of the circle is: $2\pi r = 2\pi(14) = 28\pi$

$\pi = 3 \rightarrow 28\pi = 28 \times 3 = 84$

19) Answer: C.

$(3^a)^b = 81 \rightarrow 3^{ab} = 81$

$81 = 3^4 \rightarrow 3^{ab} = 3^4 \rightarrow ab = 4$

20) Answer: C.

$\sqrt{x} = 6 \rightarrow x = 36$ then; $\sqrt{x} - 11 = \sqrt{36} - 11 = 6 - 11 = -5$

and $\sqrt{x - 11} = \sqrt{36 - 11} = \sqrt{25} = 5$

CLEP College Math Workbook

Then: $(\sqrt{x-11}) + (\sqrt{x} - 11) = 5 + (-5) = 0$

21) Answer: A.

All integers from 11 to 17 are: 11, 12, 13, 14, 15, 16, 17

The mean of these integers is: $\frac{11+12+13+14+15+16+17}{7} = \frac{98}{7} = 14$

22) Answer: C.

$-2a + 6a + 4a = 64 \rightarrow 8a = 64 \rightarrow a = \frac{64}{8} = 8$

Then; $\frac{4a-2}{5} = \frac{4(8)-2}{5} = \frac{32-2}{5} = 6$

23) Answer: B.

$|-14 - 6| - |-10 + 4| = |-20| - |-6| = 20 - 6 = 14$

24) Answer: B.

$\alpha = 180° - 132° = 48°$

$\beta = 180° - 125° = 55°$

$x + \alpha + \beta = 180° \rightarrow$

$x = 180° - 48° - 55° = 77°$

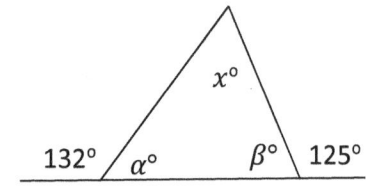

25) Answer: D.

Based on the table provided:

$g(-1) = g(x = -1) = 3$

$g(3) = g(x = 3) = -1$

$5g(-1) - 3g(3) = 5(3) - 3(-1) = 15 + 3 = 18$

26) Answer: D.

Choose a random number for a and check the options. Let a be equal to 21 which is divisible by 7, then:

A. $a - 2 = 21 - 2 = 19$ is not divisible by 3

B. $a + 2 = 21 + 2 = 23$ is not divisible by 3

C. $2a = 2 \times 21 = 42$ is divisible by 3

   but if $a = 7 \rightarrow 2a = 14$ is not divisible by 3

D. $3a - 3 = (3 \times 21) - 3 = 60$ is divisible by 3

27) Answer: B.

$(x-2)^2 = 16 \to$ Find the third root of both sides. Then:

$x - 2 = 4 \to x = 6$

$\to (x-3)(x-4) = (6-3)(6-4) = (3)(2) = 6$

28) Answer: C.

The quadrilateral is a trapezoid. Use the formula of the area of trapezoids: $Area = \frac{1}{2}h(b_1 + b_2)$

You can find the height of the trapezoid by finding the difference of the values of y for the points A and D. (or points B and C)

$h = 8 - 2 = 6$

AB$= \sqrt{(x_1-x_2)^2 + (y_1-y_2)^2} = \sqrt{(7-5)^2 + (8-8)^2} = \sqrt{4+0} = 2$

CD$= \sqrt{(x_1-x_2)^2 + (y_1-y_2)^2} = \sqrt{(8-2)^2 + (2-2)^2} = \sqrt{36+0} = 6$

Area of the trapezoid is: $\frac{1}{2}h(b_1 + b_2) = \frac{1}{2}(6)(2+6) = 24$

29) Answer: D.

First find the number of pants sold in each month.

January: 100, February: 78, March: 80, April: 60, May: 74, June: 55

Check each option provided.

A. There is a decrease from January to February

B. February and March, $\left(\frac{80-78}{80}\right) \times 100 = \frac{2}{80} \times 100 = 2.5\%$

C. There is a decrease from March to April

D. April and May: there is an increase from April to May $\left(\frac{74-60}{74}\right) \times 100 = \frac{14}{74} \times 100 = 18.92\%$

30) Answer: C.

First, order the number of shirts sold each month: 120, 130, 135, 140, 150, 160

mean is: $\frac{120+130+135+140+150+160}{6} = \frac{834}{6} = 139$

Put the number of shoes sold per month in order:

15, 20, 20, 30, 30, 35 ; median is: $\frac{20+30}{2} = 25$

31) Answer: B.

The ratio of number of pants to number of shoes in March equals $\frac{80}{35}$.

Seven-eighth of this ratio is $\left(\frac{7}{8}\right)\left(\frac{80}{35}\right)$. Now, let x be the number of shoes needed to be added in February.

$\frac{78}{30+x} = \left(\frac{7}{8}\right)\left(\frac{80}{35}\right) \to \frac{78}{30+x} = \frac{560}{280} = 2 \to 78 = 2(30+x) \to 78 = 60 + 2x$
$\to 2x = 18 \to x = 9$

32) Answer: C.

The value of y in the x-intercept of a line is zero. Then:

$y = 0 \to 4x - 4(0) = 9 \to 4x = 9 \to x = \frac{9}{4}$ then, x-intercept of the line is $\frac{9}{4}$.

33) Answer: A.

Formula of triangle area = $\frac{1}{2}$ (base × height)

Since the angles are $45°, 45°, 90°$, then this is an isosceles triangle, meaning that the base and height of the triangle are equal.

Triangle area = $\frac{1}{2}$ (base × height) = $\frac{1}{2}(6 \times 6) = 18$

34) Answer: A.

Let x be the number of years. Therefore, $2,000 per year equals $2,000x$. Starting from $24,000 annual salary means you should add that amount to $2,000x$.

Income more than that is: $I > 2,000 x + 24,000$

35) Answer: D.

The amount of money for x bookshelf is: $300x$

Then, the total cost of all bookshelves is equal to: $300x + 700$

The total cost, in dollar, per bookshelf is: $\frac{\text{Total cost}}{\text{number of items}} = \frac{300x+700}{x}$

36) Answer: A.

Number of Mathematics book: $0.20 \times 550 = 110$

Number of Chemistry book: $0.3 \times 550 = 165$

Product of number of Mathematics and number of Chemistry books: $110 \times 165 = 18,150$

37) Answer: D.

According to the chart, 50% of the books are in the Mathematics and Chemistry sections. Therefore, there are 275 books in these two sections: $0.50 \times 550 = 275$

$\gamma + \alpha = 275$, and $\gamma = \frac{2}{3}\alpha$ (Replace γ by $\frac{2}{3}\alpha$ in the first equation)

$\gamma + \alpha = 275 \rightarrow \frac{2}{3}\alpha + \alpha = 275 \rightarrow \frac{5}{3}\alpha = 275 \rightarrow$ multiply both sides by $\frac{3}{5}$: $\left(\frac{3}{5}\right)\frac{5}{3}\alpha = 275 \times \left(\frac{3}{5}\right) \rightarrow \alpha = \frac{275 \times 3}{5} = 165$

$\alpha = 165 \rightarrow \gamma = \frac{2}{3}\alpha \rightarrow \gamma = \frac{2}{3} \times 165 = 110$

There are 110 books in the Mathematics section.

38) Answer: D.

The angle γ is: $0.2 \times 360 = 72°$

The angle β is: $0.18 \times 360 = 64.3°$

39) Answer: D.

The angles on a straight line add up to 180 degrees. Then:

$x + 30 + y + 3x + y = 180$

Then, $4x + 2y = 180 - 30 \rightarrow 4(20) + 2y = 150 \rightarrow 2y = 150 - 80 = 70 \rightarrow y = 35$

40) Answer: A.

The distance of A to B on the coordinate plane is: $\sqrt{(x_1 - x_2)^2 + (y_1 - y_2)^2} = \sqrt{(8-2)^2 + (9-1)^2} = \sqrt{6^2 + 8^2} = \sqrt{36 + 64} = \sqrt{100} = 10$

The diameter of the circle is 10 and the radius of the circle is 5. Then: the circumference of the circle is: $2\pi r = 2\pi(5) = 10\pi$

41) Answer: D.

Square root of 25 is $\sqrt{25} = 5 < 6$

Square root of 4 is $\sqrt{4} = 2 < 6$

Square root of 35 is $\sqrt{35} = \sqrt{36-1} < \sqrt{36} = 6$

Square root of 49 is $\sqrt{49} = 7 > 6$

Since, $\sqrt{36} < \sqrt{49}$, then the answer is D.

42) Answer: C.

The area of the trapezoid is: $Area = \frac{1}{2}h(b_1 + b_2)$

$A = \frac{1}{2}(x)(12 + 3) \rightarrow 15x = 60 \rightarrow x = 4$

$y = \sqrt{3^2 + 4^2} = \sqrt{9 + 16} = \sqrt{25} = 5$

The perimeter of the trapezoid is: $4 + 12 + 5 + 15 = 36$

43) Answer: A.

$|x - 2| \geq 8$; Then: $x - 2 \geq 8 \rightarrow x \geq 8 + 2 \rightarrow x \geq 10$

Or, $x - 2 \leq -8 \rightarrow x \leq -8 + 2 \rightarrow x \leq -6$

Then, the solution is: $x \geq 10 \cup x \leq -6$

44) Answer: B.

Since, E is the midpoint of AB, then the area of all triangles DAE, DEF, CFE and CBE are equal.

Let x be the area of one of the triangles,

Then: $4x = 160 \rightarrow x = 40$

The area of $DEC = 2x = 2(40) = 80$

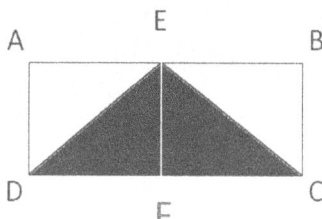

45) Answer: A.

$6 < -5x - 4 < 11 \rightarrow$ Add 4 to all sides. $6 + 4 < -5x - 4 + 4 < 11 + 4$

$\rightarrow 10 < -5x < 15 \rightarrow$ Divide all sides by -5.

(Remember that when you divide all sides of an inequality by a negative number, the inequality sing will be swapped. < becomes >)

$\frac{10}{-5} > \frac{-5x}{-5} > \frac{15}{-5} \rightarrow -2 > x > -3, or\ -3 < x < -2$

46) Answer: D.

Frist factor the function: $f(x) = x^3 + 6x^2 + 8x = x(x + 4)(x + 2)$

To find the zeros, f(x) should be zero. $f(x) = x(x + 4)(x + 2) = 0$

Therefore, the zeros are: $x = 0$, $(x + 2) = 0 \Rightarrow x = -2$; $(x + 4) = 0 \rightarrow x = -4$

CLEP College Math Workbook

47) Answer: A.

Squaring both sides of the equation gives $3m + 54 = m^2$

Subtracting both sides by $m + 54$ gives us the equation $m^2 - 3m - 54 = 0$

Here you can solve the quadratic by factoring to get $(m - 9)(m + 6) = 0$

For the expression $(m - 9)(m + 6)$ to equal zero, $m = 9$ or $m = -6$

Since m is a positive integer, 9 is the answer.

48) Answer: C.

In order to figure out what the equation of the graph is, fist find the vertex. From the graph we can determine that the vertex is at $(1,2)$. We can use vertex form to solve for the equation of this graph. Recall vertex form, $y = a(x - h)^2 + k$, where h is the x coordinate of the vertex, and k is the y coordinate of the vertex. Plugging in our values, you get $y = a(x - 1)^2 + 2$

To solve for a, we need to pick a point on the graph and plug it into the equation.

Let's pick $(-1, 10)$. $10 = a(-1 - 1)^2 + 2 \to 10 = a(-2)^2 + 2 \to 10 = 4a + 2$

$8 = 4a \to a = 2$ Now the equation is: $y = 2(x - 1)^2 + 2$

Let's expand this, $y = 2(x^2 - 2x + 1) + 2 \to y = 2x^2 - 4x + 2 + 2$

$\to y = 2x^2 - 4x + 4$

49) Answer: B.

40% of 70 equals to: $0.40 \times 70 = 28$

15% of 600 equals to: $0.15 \times 600 = 90$

40% of 70 is added to 15% of 600: $28 + 90 = 118$

50) Answer: A.

$(x^5)^{\frac{3}{10}} = x^{5 \times \frac{3}{10}} = x^{\frac{15}{10}} = x^{\frac{3}{2}} = \sqrt{x^3} = x\sqrt{x}$

51) Answer: D.

Based on triangle similarity theorem: $\frac{a}{a+b} = \frac{c}{3} \to c = \frac{3a}{a+b} = \frac{3\sqrt{2}}{\sqrt{2}+2\sqrt{2}} = 1$

$\to$ area of shaded region is: $\left(\frac{c+3}{2}\right)(b) = \frac{4}{2} \times 2\sqrt{2} = 4\sqrt{2}$

CLEP College Math Workbook

52) Answer: C.

One liter=1,000 cm^3 → 6 liters = 3,000 cm^3

$3,000 = 15 \times 4 \times h \to h = \frac{3,000}{60} = 50$ cm

53) Answer: D.

The union of A and B is: $A \cup B = \{1, 3, 5, 8, 12, 16, 24, 32\}$. There are 8 elements in $A \cup B$.

54) Answer: B.

To rewrite $\frac{2+3i}{4-2i}$ in the standard form $a + bi$, multiply the numerator and denominator of $\frac{2+3i}{4-2i}$ by the conjugate, $4 + 2i$. This gives $\left(\frac{2+3i}{4-2i}\right)\left(\frac{4+2i}{4+2i}\right) = \frac{8+4i+12i+6i^2}{4^2-(2i)^2}$. Since $i^2 = -1$, this last fraction can be rewritten as $\frac{8+4i+12i+6(-1)}{16-4(-1)} = \frac{2+16i}{20}$.

55) Answer: C.

$350,000 = 3.5 \times 10^5$

56) Answer: B.

If $\beta = 4\gamma$, then multiplying both sides by 16 gives $16\beta = 64\gamma$.

$\alpha = 2\beta$, thus $\alpha = 8\gamma$. Multiply both sides of the equation by 8 gives $8\alpha = 64\gamma$.

57) Answer: D.

To solve for the inverse function, first replace $f(x)$ with y. Then, solve the equation for x and after that replace every x with a y and replace every y with an x. Finally, replace y with $f^{-1}(x)$. $f(x) = \frac{6x-3}{5} \Rightarrow y = \frac{6x-3}{5} \Rightarrow 5y = 6x - 3 \Rightarrow 5y + 3 = 6x \Rightarrow \frac{5y+3}{6} = x; f^{-1}(x) = \frac{5x+3}{6} \Rightarrow f^{-1}(3) = \frac{5(3)+3}{6} = \frac{18}{6} = 3$

58) Answer: B.

Since a box of pen costs $3, then $3p$ Represents the cost of p boxes of pen.

Multiplying this number times 1.085 will increase the cost by the 8.5% for tax.

Then add the $8 shipping fee for the total: $1.085(3p) + 8$

59) Answer: C.

The rate of construction company = $\frac{20 \text{ cm}}{1 \text{ min}} = 20$ cm/min

Height of the wall after 30 minutes = $\frac{20 \text{ cm}}{1 \text{ min}} \times 30 \text{ min} = 600 \text{ cm}$

Let x be the height of wall, then $\frac{3}{2}x = 600 \text{ cm} \rightarrow x = \frac{2 \times 600}{3} \rightarrow x = 400 \text{ cm} = 4 \, m$

60) Answer: D.

$f(g(x)) = 3 \times (\frac{1}{x})^3 + 3 = \frac{3}{x^3} + 3$

"End"

www.ingramcontent.com/pod-product-compliance
Lightning Source LLC
Chambersburg PA
CBHW081106080526

44587CB00021B/3473